CREATIVITY AND SPACE

Ashgate Economic Geography Series

Series Editors:
Michael Taylor, Peter Nijkamp and Tom Leinbach

Innovative and stimulating, this quality series enlivens the field of economic geography and regional development, providing key volumes for academic use across a variety of disciplines. Exploring a broad range of interrelated topics, the series enhances our understanding of the dynamics of modern economies in the developed and developing countries, as well as the dynamics of transition economies. It embraces both cutting edge research monographs and strongly themed edited volumes, thus offering significant added value to the field and to the individual topics addressed.

Other titles in the series:

The Future of Europe's Rural Peripheries
Edited by Lois Labrianidis
ISBN 0 7546 4054 X

The Sharing Economy
Solidarity Networks Transfroming Globalisation
Lorna Gold
ISBN 0 7546 3345 4

China's Rural Market Development in the Reform Era
Him Chung
ISBN 0 7546 3764 6

Foreign Direct Investment and Regional Development in East Central Europe
and the Former Soviet Union
Edited by David Turnock
ISBN 0 7546 3248 2

Proximity, Distance and Diversity
Issues on Economic Interaction and Local Development
Edited by Arnoud Lagendijk and Päivi Oinas
ISBN 0 7546 4074 4

Creativity and Space
Labour and the Restructuring of the German Advertising Industry

JOACHIM THIEL
Hamburg University of Technology, Germany

Routledge
Taylor & Francis Group

LONDON AND NEW YORK

First published 2005 by Ashgate Publishing

Reissued 2019 by Routledge
2 Park Square, Milton Park, Abingdon, Oxon, OX14 4RN
52 Vanderbilt Avenue, New York, NY 10017

Routledge is an imprint of the Taylor & Francis Group, an informa business

Publisher's Note
The publisher has gone to great lengths to ensure the quality of this reprint but points out that some imperfections in the original copies may be apparent.

Disclaimer
The publisher has made every effort to trace copyright holders and welcomes correspondence from those they have been unable to contact.

A Library of Congress record exists under LC control number:

ISBN 13: 978-0-8153-8829-6 (hbk)
ISBN 13: 978-1-138-61920-3 (pbk)
ISBN 13: 978-1-351-16080-3 (ebk)

Contents

List of Figures

List of Tables

Acknowledgements

This book is the most important product of my research work with the Department of Urban and Regional Economics at the Hamburg University of Technology during the more than seven years I have spent there. Of course, as with any comparable academic work, it thus constitutes not only an individual achievement, but is essentially intertwined with the collegial and intellectual atmosphere of the place of work, with stimulating debates and discussions as well as with the direct practical support to working out this monograph.

However, first of all I am very much indebted to all the interviewees from Hamburg based advertising agencies I could talk to, for their general willingness and openness. The learning process I have experienced through these interviews constitutes the basic backbone of this book. In addition the talks completely abolished the preconceptions I previously had of people from the advertising world.

In the work context the first thanks go to Dieter Läpple as main supervisor of the underlying PhD project. His intellectual influence is visible throughout the whole book and shaped my way of thinking and arguing long before I started to work with him personally. Yet, I also owe all the other colleagues very much not only for the collegiality in the everyday work, but also for the many things I have learnt from them and for their critical comments on my thoughts. In the early periods of work this was chiefly Helmut Deecke, Birgit Kempf and Wolfram Droth, the latter above all through his enormous helpfulness and patience in teaching me the analysis of secondary data; towards the end of writing the most intensive company and discussion came from Jürgen Glaser, Jan Rathjen and Gerd Walter, Frauke Funk, Sibylle Merbitz, and in the final period of producing this monograph above all Andreas Beekmann, who supported lay out, illustration and indexing; special thanks also for that.

Outside the department the two co-supervisors of the PhD have to emphasized: Neill Marshall especially for his tip 'to go back to the beginning' which definitely put me on the right track when looking for the thread of my whole work, Thomas Malsch for quickly accepting the role as additional referee and for the appreciation of my work. In addition, more recently Oliver Ibert gave a fundamental contribution to the book, spending four and a half hours with Chapter 2. Thanks go also to Valerie Rose from Ashgate for reacting so positively to my proposal as well as to two anonymous reviewers for their affirmative comments. Despite all this backing, the responsibility for the final outcome is mine.

The support for such a project does not only comprise the academic world. I owe a lot to all persons who enabled me to go the way I have gone so far: from the very beginning my parents, and more recently Katrin particularly for her creativity in finding ways to leave me alone at home for work. It is obvious that balancing work on such a long term project with family life generated difficult situations for the whole family. That we managed to stand them is particularly due to her, and of course to Baki and Pauline who (not always easily) accepted my lack of availability. It is therefore that I dedicate this book to my family.

Lüneburg / Hamburg
December 2004

Chapter 1

Introduction

The Issue: Creativity and Contexts

'Creativity' is *en vogue*. And it is not just the rise (and fall) of the 'cool and creative' (Gill, 2002) industries like old and new media that nurtures this current excitement, but the idea of a wider transformation of the economy as a whole, including the sources of competitiveness, the organizational patterns, but also the spatial structure of economic activities. The rules of the game in the knowledge-based economy have changed: '... we now have an economy powered by creativity', as Richard Florida vigorously claims in his academic bestseller 'The Rise of the Creative Class' (Florida, 2002a).

The political and academic debates around this rise of creativity to a major asset of the knowledge economy on the one hand are very much about the individual. Creativity is a matter of 'intrinsic motivation'; it is about 'doing what you love and loving what you do' (Amabile, 1997) and about creating 'out of inner necessity' (Caves, 2000), thus being directly linked with the personal identity (Helbrecht, 1998). Creativity also reflects a specific type of behaviour; it is about the unexpected and unadapted, thus involving a sort of reformulation and generalization of Schumpeter's 'creative response' (Schumpeter, 1988a (1947)), as an action which, in contrast with the 'adaptive response', is undertaken 'outside of the range of existing practice'. And consequently, creativity is about a specific personality, the 'eccentric, alternative and bohemian' (Florida, 2002b), or, again, about a reformulation and generalization of Schumpeter's ideal type of a creative 'entrepreneur' (Schumpeter, 1988b (1949)), regarding it now as the mainstream character of a knowledge economy.

On the other hand, however, as with the role of the entrepreneur in innovation, it seems an 'illusionary idea' to conceive of creativity, regarding its importance as an input to business activities, merely as a 'unique individual quality' (Pratt, 2004). Also creative action relies on a structural context it is embedded in which can both 'enable creativity to flourish' (ibid.) and function as a constraint. Creativity thus naturally has to be read from the perspective of a 'duality of structure' (Giddens, 1984). In addition, and closely linked to that, also the growing importance of creativity in economic action must be understood in the wider context of structural change. Three main fields – and their corresponding debates – have to be emphasized which may frame the increasing demand for more creative individuals in the contemporary economy:

First, even though the new hype around creativity is based on a broad idea of socio-economic change it has to be discussed in view of the rising 'creative' or 'cultural industries' which are said to constitute somehow the institutional and organizational role model for the new creative economy (Thompson and Warhurst, 2004). The main point within the debates around creative industries is that there is an increasing convergence of the formerly separated spheres of culture and economy, with this argument drawing on the pioneering arguments of the Frankfurt School (e.g. Adorno, 1991), however coming to converse conclusions. Instead of identifying an increasing subordination of culture to the logic of capitalist mass production, many contemporary advocates of 'creative industries' hold to recognize a 'cultural turn' of the economy (Crang, 1997), thus focusing rather on a subordination of the economy to a cultural logic (Lash and Urry, 1994). Notwithstanding the questions about the validity of this position the approximation of two areas of society based on different 'frameworks of action' (Storper, 1997) has implications for the practice of economic life given that these different frameworks also involve production processes, organizational settings, 'performance criteria' (Girard and Stark, 2002) etc. Thus a growing degree of creativity may also comprise changes in the field of *economic organization* in that patterns inherent to cultural production encounter the 'classic' organizational and institutional setting of a capitalist market economy.

Second, the way the individual as a creative subject is conceptualized points to a generally wider understanding of economic action beyond the atomized, fully informed and rationally acting 'homo oeconomicus'. The fact that creative action involves the very personality – and not just abstract skills – implies that, with an increasing importance of creativity, also the social basis of this personality increasingly contributes to its economic or professional activity. This socially informed perspective of the economy is even more valid when looking at the narrower field of the creative industries: As cultural production strongly depends 'on inter-personal norms, methods, languages and so on, in order to achieve communicability' (Scott, 1999a, p.808), a high degree of social underpinning appears to be a necessary feature of the creative economy also from an output-perspective. That is to say, dealing with creativity requires dealing with the wider debates on the recontextualization of the economy in a post-industrial era, respectively on its embeddedness in networks of *social relations* (Granovetter, 1985), dominating big parts of industrial sociology, innovation economics, regional science and economic geography etc. during the last about 30 years.

Third, one important feature of the debates on recontextualization has always been its spatial dimension, based on the fact that even in an increasingly globally connected world economies of localization and urbanization, that is, forms of economic success based on short-distance relations, do not only survive but seem to increase in importance. The creativity discussions tend to accentuate this idea of a 'compulsion of proximity' (Boden and Molotch, 1994) even more strongly. As with the recontextualization this has two dimensions: On the one hand, the social underpinning of creative activities is manifested in space, given a spatially selective crystallization of 'inter-personal norms', conventions and institutions in the related industries (Scott, 1997, 1999a; Cristopherson and Storper, 1986). On the other hand, the socio-spatial environment, which involves social relations, institutional settings, but also the materially built environment, can both support and hinder creative action (Helbrecht, 1998; Florida, 2002a; Drake, 2003). That is to say, the rising importance of creativity closely interacts with the *spatial structure* of the economy, shaping it by reconfiguring the pattern of prosperity and decline, in turn however also being shaped by a variety of socio-spatial contexts influencing individual creative action in different ways.

In sum, thus, discussing the rise of creativity as a characteristic trait of the contemporary knowledge-based economy requires approaching this economy in a multifaceted way: neither focusing on the capabilities of single creative individuals nor simply stressing a 'duality of structure' leads to the point given that the structural context of the knowledge-economy is more complex than the corporation-centred pattern of the post-war era (Chandler, 1977). Dealing with a creative economy thus necessarily implies dealing with the intricate interplay between increasingly individualist action and the context it is embedded in, but also taking into account that this context is shaped by the interrelations between *economic organization*, *social relations* and *spatial structures*.

The Arguments: Labour Markets and Space

This book is precisely about the outlined complex duality of individualism and context, focusing on the seeming paradox that the current structural change towards a knowledge-based economy fosters both an 'individualization' of the economy, that is, a 'disembedding' of individuals from traditional structures, and the growing importance of its social and geographical underpinning, that is, their obvious 're-embedding' in new socio-spatial configurations (Giddens, 1990). Thereby its approach on the one hand substantially differs from the neoclassical accounts of the economy based on the 'atomized' actor and the assumption of a market obstructing effect of underlying social mechanisms. On the other hand, it argues also against the mainstream work on the socio-spatial embeddedness of economic activities having come up in the last 25 years and emphasizing the concurrence of economic specialization, socio-cultural inclusiveness and spatial proximity as the new prime model of production organization in a post-industrial or post-Fordist world (Piore and Sabel, 1984; Sabel, 1989, among others). Obviously the interwovenness of *economic organization*, *social relations* and *spatial structures* both functions in a more complex manner and it interacts with a secular process of social individualization, that is, it is driven and shaped by increasingly 'reflexive' individuals acting independently from given structures both regarding their work context and their lives (Giddens, 1990, 1991; Lash and Urry, 1994).

This given, the first basic argument underlying the following chapters is that it is essentially the labour market of 'post-industrial professionals' (Bell, 1974) through which the intricate interaction of increasingly individualist reflexive action with its different economic, social and spatial contexts is shaped. The labour market of reflexive activities thus constitutes so to speak the key 'arena' of the knowledge-based economy and the key 'anchor' which makes contemporary economic action and organization increasingly interwoven with its social and spatial environment. This is not to say, the economy merely 'obeys' the growing importance of labour as an input to production in a straightforward way. First, it is rather the concurrence and interaction of strategies undertaken by knowledge firms and knowledge workers on the labour market, regarding the optimization of quality and cost of production, the successful recruitment of labour, but also regarding biographical plans and professional identities which drives the organization and structure of knowledge-intensive activities. Second, and more importantly, this complex interaction and the increasing involvement of the whole person in professional activities do not only raise the importance of labour, but also the uncertainty regarding its actual performance. This ambivalence of importance and uncertainty is driven by the nature of labour as a 'fictitious commodity' (Polanyi, 2001 (1944)), that is, by the fact that labour is not produced for being sold on a market but actually consists of human beings within the relational contexts of their individual biographies.

The second argument takes up this fundamental ambivalence characteristic for creative knowledge labour, holding that it is precisely the uncertain nature of labour markets its social and, even more importantly, its spatial underpinning is based on. Unlike other accounts emphasizing firm location and spatial development as being based on good opportunities or 'factor endowments' we thus consider the spatial context of the knowledge-based creative economy, at least to a significant extent, as a means to support the diverse economic actors' strategies of dealing with the uncertainties, respectively with the ambivalences inherent to an increasing labour orientation of the economy. And it seems to be that urban or metropolitan environments with a sufficient degree of socio-cultural diversity prove most suitable to provide this support. Without aiming to get into any determinism as to a particular spatial form of the economy we do thus conclude arguing that an increasingly creative knowledge economy tends to be also an increasingly urban economy (Glaeser, 1998).

These general arguments are to be substantiated and illustrated by an in-depth empirical study of advertising as a service industry on the one hand paradigmatic as to the encounter of different economic 'action frameworks' and organizational patterns, the embeddedness of professional activities in dense networks of social relations, and its spatial concentration and inherent urbanness. On the other hand these different contextual fields are shaped by and interact with an exceptionally reflexive workforce of creative professionals essentially having to generate output 'outside of the range of the existing practice' in order to attract consumers' increasingly scarce 'attention'.[1] Given that the advertising labour market thereby exemplarily reflects the intricate interaction of creativity with its contexts it is likely to anticipate future patterns of the economy particularly well.

The Structure of the Book

The bulk of this monograph consists of the empirical study about the restructuring of the German advertising industry during the last about 25 years, focusing above all on the rise of Hamburg as the country's new 'creative capital' as well as on the role of labour market mechanisms as catalysts of this change. Three out of the six chapters present the results of both extensive secondary data research and intensive fieldwork including about 30 semi-structured interviews and thorough documentary analysis having been carried out since 2000, including a concentrated work period in the years 2000 and 2001. Yet, as might have become clear in the preceding introductory sections, the book aims to offer more than just a case study of a particular business service, intending to contribute to the current debates on socio-economic change by stressing the critical role of the labour market both as 'arena' of this change and as 'medium' of its socio-spatial underpinning.

The structure of the book reflects this double objective of thoroughly examining a particular industry and sustaining thereby a more general argument. Chapter 2 elaborates the outlined concept of the recontextualization of economic activities in the post-industrial economy as an intensified interaction of economic organization, social relations and spatial structures in more detail. The approach is developed by means of three major academic debates of the last 25 years having provided clues as to an understanding of how post-industrial economic activity interacts with its social and spatial context: the discussion on 'new regionalism', trying to explain the surprising success of specialized firm clusters in previously rural or semi-rural regions, the 'new centrality' debate stressing the persistent importance of agglomeration economies in a globalized economy, and the recent work on the 'creative economy' mentioned above. The arguments put forward in these debates are confronted with a 'subject-centred' concept of recontextualization, drawing mainly on Giddens' (1990, 1991) as well as Lash and Urry's (1994) arguments about an increasing need for 'reflexive' individual action, implying a larger degree of freedom, yet also of uncertainty, for individuals in both work and life of modern societies, thereby bringing about a 'convergence' of both spheres. This 'convergence of work and life' in turn however tends to accentuate the 'commodity fiction' characteristic for labour (Polanyi, 2001 (1944)) given that it also approximates its two contradictory dimensions as commodity and human being. It is this 'conflicting convergence' our key claim is based on: that the labour market of reflexive activities, respectively the process of filtering the

---

1   See also Chapter 3, p.46.

convergence in this labour market, essentially shapes the functioning of the post-industrial economy.

In Chapters 3 to 5 the results of the empirical work are unfolded from three different perspectives, in the course of which the analysis is increasingly directed towards the 'subjects' within the advertising industry. Chapter 3 provides a more general overview, at first of advertising as an economic activity and the general structural change the industry has undergone since the end of the 1970s, generally labelled as the transition from the 'first wave' to the 'second wave' of global advertising, and subsequently outlining the peculiarities of the German advertising market and the territorial structure of its production system. The chapter chiefly concentrates on how the general change of global advertising is reflected in the German space-economy, emphasizing the shift to the favour of Hamburg and offering a first explanation of this shift: On the one hand it is derived from the changes in global advertising towards a more creative 'second wave' style which, on the other hand, could be adapted to the German market by a group of Hamburg based pioneer agencies. Drawing on Storper and Walker (1989) we conceive of the performance of these pioneers as a process of innovation in which the leading actors are provided with a high degree of 'locational freedom', being likely to locate outside existing centres and to 'produce' their own regions.

Chapter 4 focuses on the firm, respectively on advertising agencies as key actors of the national industry, thereby examining this innovation process in a more exhaustive way. The 'substance' of innovation is presented by portraying the story of the most significant pioneer agency, Springer & Jacoby, which managed to create a new product, that is a more entertaining advertising style, by decentralizing the organization of the firm thereby both enhancing individual responsibility and mobilizing the internal labour market. Subsequently the general impact of this 'entrepreneurial' innovation on the German landscape of advertising agencies is discussed, outlining the changing 'action frameworks' of five different types of agencies varying in market size and innovativeness which make up this landscape. The chapter concludes with a comprehensive interpretation of the innovation process, launching the argument that the change within the advertising industry can be conceived of as a mutual bridging of two separate worlds it consists of: the world of business services and the world of popular arts. The innovative pioneers contributed to this bridging by both adding to the importance of the popular arts world and proving to be successful even in terms of the business performance criteria. Thereby they tended to shift the key arena of competition in the industry to the labour market of creative professionals.

In Chapter 5 this labour market is taken under closer scrutiny, in two steps: First, we discuss it as the key to understanding the innovation process. Enhancing the importance of creativity presupposed getting access to artistic segments of the labour market; channelling the creativity into the business service world in turn presupposed adapting artistic labour to the requirements of a business service. Second, from the perspective of the labour market also the present structure and the continuous restructuring of the industry can be understood. Basically, the interaction of an increasingly globalized service industry with an increasingly individualist labour force adds to the volatility of both the labour market, characterized by extremely high turnover rates, and the agency landscape, being constantly restructured by mergers and buy-outs, but also by the continuous emergence of new agency start-ups. In a final interpretation of the book's empirical part we take up again the idea of the 'locational freedom' for leading firms in a process of change and discuss it in view of the findings. We hold that in an innovation process essentially involving creative labour the freedom is only relative since the location has to be oriented to the need for accessing or attracting labour. In this sense Hamburg's rise as Germany's 'creative capital' can be seen as a combination of the pioneer's strategies and the fact that the city provided both an attractive environment and the necessary urban 'massivity' to appeal to potential creative labour force.

The final Chapter (6) aims to grasp the spatiality of an economy relying essentially on reflexive knowledge labour in general conceptual terms. Drawing on a critique of the interpretations of the labour factor in classic work on regional restructuring the argument focuses the 'ambivalence of importance and uncertainty' as the fundamental feature of reflexive labour as key input into the knowledge-based economy. This ambivalence is reflected in the basic 'trinity' of economic organization, social relations and spatial structures which constitutes a sort of latent analytical framework of the whole book. The chapter concludes holding that there are many reasons pointing to an increasing importance of urbanization economics, given that a 'thick' and diverse urban environment helps economic actors to deal with the inherent ambivalence of reflexive labour markets. Generally put, Chapter 6 stands paradigmatically for perhaps the most ambitious objective of this book, that is, to divulge the idea of an inherent spatiality of the future knowledge-based economy, based on the complex mechanisms at work in the markets of the key 'fuel' of this economy: human labour.

Chapter 2

The Economic, the Social and the Spatial: Framing the Post-Industrial Economy

'The Economy on its Way Back ...'[1] – Recontextualizing Economic Action

It was almost thirty years ago that Daniel Bell, in his famous and ambitious 'venture in social forecasting', outlined the contours of a 'post-industrial society' (Bell, 1974) in which the 'primary logic' of socio-economic organization would shift from the 'co-ordination of machines and men for the production of goods' in industrial society to being 'organized around knowledge'. His work, despite being strongly criticized above all for the too optimistic view of post-industrial employment in terms of both quantity and quality and due to a too 'mechanistic' understanding of the role of knowledge for the economy, marks the beginning of extensive academic work on the knowledge and information economy, involving a whole variety of disciplines and perspectives. This chapter does not aim to provide an exhaustive account of the 'polyphony of voices' (Amin and Cohendet, 2004) around knowledge; it focuses on one particular aspect lying at the heart of Bell's argument without being explicitly elaborated which has strongly influenced the academic debates of the last 30 years: The post-industrial economy has an inherently social dimension based on the fact that knowledge is 'embodied in institutions and represented by persons' (Bell, 1974, p.212), thereby implying a new class structure, with a new group of professional, technical and managerial workers as both source and outcome of a knowledge economy, as well as a new mode of interaction, less between 'machines and men' than based on interpersonal relations. Or, in the author's words (p.488): 'The post-industrial society is essentially a game between persons.'

Particularly the latter suggestion of the economy's increasing interwovenness with the social relations it is embedded in is remarkable, given that it completely disagrees with the classic views of the relation between economy and society (not only) inherent to economic thinking both in a static and in a dynamic perspective.

Regarding the former it challenges the analytical concept of the 'homo oeconomicus' as key for the understanding of economic action in the neoclassical

---

1   'The economy on its way back into society' is the subtitle of the German translation of Piore and Sabel's landmark book 'The Second Industrial Divide' (1984).

paradigm of contemporary economics, contrasting it with a more encompassing contextualized view of the economy, considering the individual's motives and values as well as the conditions of their production and reproduction (Etzioni, 1988), his or her embeddedness in ongoing networks of social relations (Granovetter, 1985) as well as the role and functioning of 'economic institutions as social constructions' (Swedberg and Granovetter, 1992).

As to the dynamic perspective it questions the classic views of Adam Smith, for instance, conceiving of modernization as the lifting of the economic out of its social environment and the subordination of the latter to the former thereby allowing the market forces to unfold their wealth maximizing capacities. Yet this view is not only characteristic for liberal economics. Also critical social science conceived of capitalist modernization as a process of 'disembedding' the economy from its social foundations through the emergence of a market economy (Polanyi, 2001 (1944))[2] or as a 'decoupling of systems world and life world' (Habermas, 1981). Thus, seeing the post-industrial economy increasingly as a 'game between persons' basically implies an apparent turnaround in the direction of socio-economic modernization. The economy is, as the German translator of Piore and Sabel's pioneering work on 'the second industrial divide' interprets the authors' key argument, 'on its way back into society' (Piore and Sabel, 1984).

As put initially, the issue of an increasing degree of 'social foundation' of the economy in a post-industrial era has been debated most strongly in relation with the spatial structure of economic activities. The overall changes in the territorial pattern of the Western industrial economy from the crisis in the mid 1970s onwards, characterized by the transition from the mass-production led post-war model of urban growth diffusing into the rest of each national space-economy to a more diverse constellation of growth and decline of both urban and non-urban spaces had triggered a re-thinking of the basis of regional economic performance in an era after industrial mass-production stressing above all the importance of seemingly non-economic factors which lie at the heart of the economic success of cities and regions. The concept of atomized economic actors merely linked by simple market transactions was not able to explain the surprising success of specialized regionally clustered high technology production systems like Silicon Valley or the industrial districts of the Third Italy (Grabher, 1993a), the latter constituting the role model for future economic organization in the work of Piore and Sabel, respectively of the whole industrial districts-school having risen around them. The economy's way thus seemed not only lead 'back into society' but also 'back into the region' thereby also questioning another facet of the process of modernization as a 'lifting of social relations out of their localized contexts of interaction' (Giddens, 1990).

---

2   Yet, one must consider that Polanyi insists on the fact that 'the idea of a self-adjusting market implied a stark utopia. Such an institution could not exist for any length of time without annihilating the human and natural substance of society' (Polanyi, 2001 (1944), p.3). See the detailed discussion of Polanyi's arguments on p.34f.

That is to say, the post-industrial knowledge economy empirically reveals an intense interaction with the social relations and the spatial structures economic activities are embedded in. As pointed out above, this basic finding of academic work in economic sociology, evolutionary economics, organization science, urban and regional studies etc. during the last 25 years constitutes the whole book's point of departure. This chapter on the one hand briefly summarizes the main arguments conceptually grasping this interaction within the corresponding debates. On the other hand it involves a fundamental critique of these arguments based on the contention that they have not managed to offer a conclusive explanation of why it is presently, at the outset of the post-industrial knowledge economy, that the process of socio-economic modernization seems to be turned upside down.

This given, the argument of the chapter runs as follows: In the next section the existing discussions on how the post-industrial economy is interwoven with its social and spatial environment will be summarized by means of three exemplary debates from a heterodox disciplinary context, yet all with strong linkages to urban and regional studies. Subsequently the positions regarding the logic of the socio-spatial foundation of economic activities developed in these debates will be confronted with an approach which stresses the structural change lying at the heart of the 're-embedding' dynamics of the contemporary economy as an essentially social process. Drawing on the work of Giddens (1990, 1991) as well as Lash and Urry (1994) the key argument is that late modern 'reflexive subjects' both in their individual biography and in professional life are ever less able to behave based on a pre-given and self explaining set of rules. This involves both an ambivalence between increased freedom and increased uncertainty in both spheres and an intricate 'conflicting convergence' of them which is filtered through the labour market. The next section discusses precisely the logic of the labour market drawing on Polanyi's concept of labour as a 'fictitious commodity' and trying to extrapolate his arguments to the reflexivity debate. As a conclusion we shall provide a summary of the main arguments, finally claiming for an approach to recontextualization based on reflexive actors and their interaction with the functional-structural logic of the contemporary economy.

Economic Organization, Social Relations and Spatial Structures: Three Examples

It would be misleading to argue that societal modernization since the industrial revolution in the 19th century – and the academic debate on it – has consisted in or has been conceived of as, respectively, a linear process of 'space-time distanciation' (Giddens, 1990) and social disembedding. Even urban and regional economics since its very beginning, overlapping precisely with the growth of the industrial city at that time, has considered short distance relations between economic actors to be important, also, and particularly within a modern economy. Alfred Weber's concept of *agglomeration economies*, broken down into *economies of localization* as the cost savings through agglomeration of firms from one sector and *economies of urbanization* as the savings obtained via spatial concentration of firms from many sectors, has always been 'one of the most powerful notions in regional economics and economic geography' (Moulaert and Djellal, 1995).

In addition Weber – in his largely ignored 'capitalist' or 'realist theory of industrial location' (Weber, 1923) – offered some hints as to the social underpinning of agglomeration by interestingly seeing it as chiefly based on the 'labour substratum' of the industry, caused by the transformation of 'labour force into a free commodity' and the 'laws of flow and motion' bringing people into the urban labour markets.[3]

Nevertheless, while agglomeration economies as a technical term has survived also in classic regional economic thinking, it has never become an integral element of economic theory, being dealt with as a sort of theoretical 'dummy' allowing to maintain the spatial dimension as external to the economic core mechanisms (Läpple, 1991). Moreover, particularly in the post-war 'Fordist' period of modernization the powerful institutional dualism of market economy and welfare state seemed to supersede the social and spatial pattern economic development is embedded in.

With the 1970s crisis this appears to have changed substantially, visible also in the way at least parts of academia have dealt with the processes of transformation the economy has been undergoing since then. In the following paragraphs we shall therefore outline three important debates of the last 25 years, linked to urban and regional studies but not only influenced by them, in chronological order which stress in different ways the increasing interaction of the economy with its socio-spatial context as essential element of major processes of socio-economic change: first, the *new regionalism*, orbiting around the rise of specialized regional economies to successful 'clusters'; second, the *new centrality*, focusing on the increase of urbanization economies in the light of globalized economic activities; third, the 'culturalization' of the economy and the assumed increasing significance of place as an environment in which a *creative economy* can unfold.

New Regionalism: Flexible and/or Innovative Clusters

The most remarkable and at that time also surprising example of a seeming basic change in the direction of economic development was the rise of a group of formerly either completely rural or craft-dominated areas to successful specialized regional economies. The success stories of Silicon Valley and others brought about two key assumptions: empirically that these regions are better adapted to the increasingly turbulent environment of the world-economy and, theoretically that, in such a context, the region generally gains importance as a level on which the foundations of economic success or failure are laid (above all in comparison with

---

3 For a more comprehensive discussion of Weber's 'capitalist theory' and the seeming ignorance related to it in the academic community see Läpple (1991).

the national state) signalling a more fundamental change in capitalist production. According to Michael Storper's summarizing view (1997, p.3) the findings suggested 'that there might be something fundamental that linked late 20th-century capitalism to regionalism and regionalization', based on the fact 'that the matrix of production [...] was an area, not a firm' (Sabel, 1989, p.17).

This 'new regionalism' strongly resembled what Alfred Marshall in the beginning of the 20th century had labelled as 'industrial districts' in order to describe the patterns of early 'pre-factory' (Amin, 2000) industrial organization. That is to say, it did not only thrive on hard external economies of localization according to Marshall based on 'labour pooling' and the 'provision of intermediate inputs', which Storper referred to as 'traded interdependencies' (Storper, 1997a) but was underpinned by a sense of commonness, an 'atmosphere' in the local or regional society which – again in the Marshallian original – above all facilitated knowledge spillovers:

> The mysteries of the trade become no mystery; but are as it were in the air. [...] Good work is rightly appreciated, inventions and improvements in machinery, in processes and the general organisation of the business have their merits promptly discussed: if one man starts a new idea, it is taken up by others and combined with suggestions of their own; and thus it becomes the source of further new ideas (Marshall, 1920, quoted from Krugman, 1991, p.37f.).

The protagonists of new regionalism developed an even broader and more general perspective of the 'untraded' nature of externalities, which regional success stories are based on, beyond the feature of facilitated knowledge spillovers, characterizing them as 'a relatively homogeneous system of values and views, [...] a system of institutions and rules [...] to spread those values throughout the district, to support and transmit them through generations' (Becattini, 1990). In other words: the social practices and cultural conventions of a regional society provide economic actors with an enhanced degree of social capital and/or trust that underpins the competitiveness of regional enterprises thereby strengthening the economy of the whole region.

The empirical cases basically orbited around two distinct types of specialized regional clusters: craft-dominated networks of diffuse rural industrialization in which traditional industries flexibly specialize either in quality-based niche markets, as in the case of the industrial districts of the Third Italy, or along the supply chain of large corporations – paradigmatically visible in the car industry of Baden-Württemberg (Amin, 2000), and high-technology districts emerging as 'new industrial spaces' (Scott, 1988a) in a completely rural and agrarian environment outside the previous centres of the economy – epitomized by Silicon Valley, the M4-corridor and others. These two types do not only reflect differences in terms of their economic history and sector composition; they also symbolize different ways of conceptualizing the substance of how the interaction of economic organization, social relations and spatial structures works, respectively how it underpins economic development, thus representing two different dimensions of a new regionalism. The first basically involves the aspect of 'flexibility' in a turbulent and fragmented market environment; the second dimension is based on the issues of 'innovation' and 'learning'. Both dimensions are closely interrelated and most of the literature of new regionalism tackles both; nonetheless they obey to different basic logics of how spatial agglomeration and social underpinning of economic activities shapes the performance of regions.

Flexibility The starting point of the 'flexibility'-dimension was the breakdown of the stable mass markets of the post war period due to both an increasing fragmentation and volatility of consumption patterns, thereby also implying the crisis of the big vertically integrated firm and thus bringing about a general decentralization of industrial production. As Charles Sabel put it:

> ... in the early 1970s, as international competition increased and world markets fragmented, firms became more and more wary of long-term investment in product-specific machinery. The product's market often disappeared before the machinery's costs were recovered. The more volatile markets became, the more firms experimented with flexible forms of organisation which permitted rapid shifts in output. As they did, they encouraged the reconsolidation of the region as an integrated unit of production (Sabel, 1989, p.18).

The nexus between increasingly decentralized and flexible forms of production organization and regionalization was most explicitly put in the work of the Californian School of Economic Geography (see Scott, 1984a, 1988a, 1988b; Storper and Christopherson, 1987; Storper and Walker, 1989, among others). Drawing basically on Coase's and Willamson's theoretical edifice of transaction costs they developed an 'economic model of the agglomeration process' (Storper, 1995) in which firms in the face of uncertainty, in order to avoid the 'misallocation of resources', tend to vertically disintegrate the production process, respectively to 'externalize much of the uncertainty by engaging in subcontracting activity' (Scott, 1988b, p.37). This vertical logic of splitting the process of production into small units of activity is complemented by a logic of horizontal disintegration 'if internal economies of scale within these units are limited' (ibid., p.52). However, the process of fragmentation in turn increases transaction costs thereby equally increasing the 'inducement for producers to locate in close proximity to one another' (ibid.). This proclivity to agglomeration is the more likely the more complex a transaction is:

> In very general terms, the greater the substantive complexity, irregularity, uncertainty and uncodifiability of transactions, the greater their sensitivity to geographical distance. In all these circumstances, the cost of covering distance will rise dramatically and in some case, no matter what the cost, the transaction becomes unfeasible at great distance. In contrast, the less frequent a set of transactions, or the more substantively simple, codifiable, certain, predictable or regular they are, the less they tend to be hampered by geographical distance (Storper and Scott, 1995, p.506f.).

As a result of this path 'from disintegration to agglomeration' (Scott, 1988b, p.52) regionally clustered 'quasi-firm' production networks emerge, specialized as a region along a particular value chain but able to flexibly adapt to market changes both quantitatively and qualitatively through being specialized as single firms in one particular step within the value chain, respectively through the economies of scope and variety thereby achieved in the region. In Piore and Sabel's 'second industrial divide' – as in nearly all the literature on the industrial districts of the Third Italy – this logic of clustering was termed as 'flexible specialization'. In the historical case of the Italian districts it was underpinned by further technological and organizational innovations in the production chain directed to increasing the versatility. The Benetton case of, on the one hand, linking point of sale and production by information technology based systems of distribution logistics in order to quickly transmit the information of a changing consumer demand to the shop floor and, on the other, of assembling knitwear without dyeing it before, in order to be able to react to that, constitutes a paradigmatic narrative of this flexibility (see Knox and Agnew, 1994). Sabel (1989) even holds that flexible specialization inverts the principles of mass production by constituting 'the manufacture of specialized goods by means of general purpose resources rather than vice-versa'.

Thus increasing flexibility does not only involve changes in the pattern of division of labour within and between firms as the Californian School's agglomeration model suggested. In order to manage an inherent trade-off between flexibility and specialization it entails technological and organizational changes and, above all, it raises the demands regarding the capabilities and the versatility of the labour force. As Capecchi put it: 'The key to the system of flexible specialization is the middle skilled group who come to the factory as professionally skilled workers and, while on the job, increase their professionality with experience' (1990, p.21).

Seen in terms of how the spatial and social context shapes regional economic performance one can hold that above all the increased complexity of a flexible production organization requires to be compensated by socio-spatial foundation. The fragmented pattern of production organization increases transaction costs; the intricate coexistence of 'cooperation' and 'competition' in the networks of production stressed by nearly all commentators in the new regionalism-literature presupposes an enhanced degree of trust between regional actors which is not generated by the very aims of individual economic activity alone but has to be produced externally. Yet, once a regional production network is established it serves as an own trust generating entity spilling over into the social and spatial context. Thus there is a positive feedback circle at work between the pattern of production organization and its socio-spatial underpinning, pointing to the fact that it is more than the mere capacity of flexible adaptation to changing markets that makes up the concurrence of economic specialization, social inclusiveness and spatial proximity characteristic for the successful cases of new regionalism.

Innovation and Learning Critics of the transaction cost based agglomeration model and the flexible specialization literature held that both, given their focus on adaptation to externally induced changes, 'consider the local relationships mainly in terms of locational efficiency, and therefore within a static approach to the interpretation of economic space' (Camagni, 1991a, p.2). Thus they stress too little the learning capacity of regional production systems which Sabel (1989) summarizes by describing flexible specialization as '... a system in which firms know that they do not know precisely what they will have to produce and further that they must count on the collaboration of workers and subcontractors to meet the market's eventual demand' (Sabel, 1989, p.53).

In this view learning within a regional cluster of specialized small firms can be seen as being based on risk-sharing and on the Marshallian knowledge spillovers pointed out above: Risk sharing mainly functions via co-operative investment in uncertain fields, and the spillovers work through the rapid diffusion of technological know-how, respectively its adjustment to the practical needs of industrial production, be it via a fluid labour market which allows workers to frequently change between firms or between waged work and self-employment (Capecchi, 1990; Becattini, 1990), via information exchange within social institutions (Asheim, 1992), via cooperation within supplier networks or among equal firms (Brusco, 1990) or even via competition among them (ibid.).

More explicitly dynamic approaches to economic space are based on the 'new evolutionary paradigm', thus drawing on the work of a 'neo-Schumpeterian' school of economic thinking, represented by David (1985), Arthur (1994), Nelson and Winter (1982) and others and basically orbiting around the issue of how (and where) technological innovation occurs. Notwithstanding the diversity of issues and arguments elaborated by the various scholars one can roughly point out three key propositions from which their relation to new regionalism might be deduced:

1. Innovation is highly uncertain, thus both consisting in an interactive process (i.e. not in a straightforward manner from research to practice) and being based on, respectively 'shaped by a variety of institutional routines and social conventions' (Morgan, 1995).
2. The innovation process is subject to increasing returns to scale and thus tends to be path-dependent. The famous example of the QWERTY-keyboard (David, 1985), which while having originally been designed to avoid striking two contiguous keys consecutively is nowadays suboptimal but still prevailing, shows that technologies are 'products of interdependent choices' (Storper, 1995). In other words, the innovation and diffusion of new technologies depends on decisions and actions which in turn are consequent upon and shaped by previous ones thereby not automatically leading to the economically most efficient solutions. This in turn on one hand increases uncertainty given the lack of knowledge about the impact of future choices; on the other hand it tends to lock technological development into certain trajectories once (more or less) accidentally chosen.

3. The path-dependent nature characteristic for technological development also
    applies to the institutional and social base of innovation pointed out above.
    Also social systems tend to be locked in trajectories of institutions and
    conventions detached from actual interaction between persons. According to
    David (1994) institutions can therefore be characterized as 'carriers of history'.

This given, 'evolutionary thinking' – according to Michael Storper (1997) –
'can be applied to problems of regional development in two ways', that is,
regarding the way how regional institutional and conventional settings contribute to
technological trajectories and in terms of the evolutionary 'behaviour' of these
settings themselves. The classic new regionalism literature tended to stress the first
aspect, focusing on where the successful development of technologies took place
and on whether and how the socio-spatial context contributed to this. There were
two contrasting positions of how the interaction between innovation through
innovative actors and the environment they are embedded in works, either as an
enabling environment, an incubator for innovative action, or as an obstacle
innovative actors have to escape from. Interestingly, this latter focused precisely on
the very evolutionary behaviour of regional institutional settings.

The first position was represented above all by the GREMI-school of
'innovative milieus' (Aydalot and Keeble, 1988; Camagni, 1991a, 1991b),
conceiving of the local environment as driver of firms' innovative capabilities.
Studying the phenomenon that high technology industry growth occurred only in a
restricted set of localities and regions, the local environment was considered to be
the 'best angle to attack these changes' (Aydalot and Keeble 1988: 8). Or, more
precisely:

> The local environment-based approach is arguably the most fruitful. Its central concern
> is to understand the firm in its local and regional context and to ascertain what
> conditions external to the enterprise are necessary both for the creation of new firms and
> the adoption of innovations by existing ones. The firm, and the innovating firm, are not
> viewed as pre-existing in or separate from the local environment, but as being products
> of it. Local milieus are regarded as the nurseries, the incubators, of innovations and
> innovative firms. [...] This line of argument leads naturally to the hypothesis that it is
> often the local environment which is, in effect, the entrepreneur and innovator, rather
> than the firm (ibid., p.8f.).

That is to say, the milieu is seen as a sort of local 'synergy space' (Camagni
1991b: 136) in which complexity and uncertainty are reduced due to a common
understanding of decision routines, a rapid information flow, risk sharing etc.,
thereby boosting the local production system's capacity of collective learning and
innovation.

Läpple (1994) conceives of the main mechanism at work within the milieu as a
'filter function' characterized by a certain socio-cultural disassociation from,
respectively closure to the external world. He argues, however, that this function
does by no means automatically imply enhanced innovative capability. Drawing on

the experience from the German Ruhr region he contrasts the 'innovative milieu' concept with the notion of a 'sclerotic milieu' (ibid.) acting as a filter that hinders the region to innovate[4] through being 'locked in' the 'strong ties' of a previously successful development trajectory (Grabher, 1993b).

The protagonists of the second position were members of the above mentioned Californian School of Economic Geography who – besides looking at the vertical disintegration of traditional industries – tried to explain the emergence of the 'new industrial spaces' linked to the information technology revolution. The key concept within their approach is the 'window of locational opportunity', corresponding to a sort of time slot during which innovative firms are basically free in terms of their locational decision. Or, in the words of Storper and Walker:

> ...there is ample reason to believe that leading firms in a rising industry do not face severe locational specification constraints attributable to needs for labour, resource inputs or inter-industrial linkages for manufactured goods, because innovation necessarily means solving technical problems presented by new ways of producing, organizational problems of how to secure (or produce) various inputs, and labour problems of mobilizing and training workers. In sum, growing industries may be said to enjoy – for a certain period [...] – both a factor-creating and factor-attracting power. [...] These moments of enhanced locational freedom may be called *windows of locational opportunity* (Storper and Walker, 1989, p.74f., *original emphasis*).

Innovation processes are thus held to be initiated by small groups of 'first movers' largely independent of their local environment and even tending to flee from it; innovative regions are considered to be produced be these innovative actors. However, Storper and Walker do not provide any conclusive clue for conceptualizing the link between these first movers and the emergence of a growth centre based on their activity.

The prime study case for an innovative high-technology region has always been the Silicon Valley in Northern California, and its localization outside the traditional urban cores was the main starting point for the idea of an enhanced locational freedom. More detailed empirical accounts of the region's recent economic history (Saxenian, 1994a, 1994b) do not stress either an external or an internal logic as main driving force of innovation. Instead innovation is held to be triggered by a combination of single actors 'unusually open to risktaking and experimentation' for 'having left behind families, friends and established communities' and a 'collective innovation' process based on 'dense social networks and open labour markets' within the region (Saxenian, 1994b). Also the continuity of Silicon Valley's innovativeness primarily relies on this dialectic of openness and closure, recently

---

4 Also the GREMI work, at least of the second generation, considers the possibility of decreasing innovativeness of milieus. Camagni uses a metaphor from thermodynamics for this, arguing that a closed system over time runs the risk to die the 'entropic death', therefore requiring 'injections of external energy' (Camagni, 1994, p.84). To guarantee these injections firms within milieus indispensably need the contact to wider networks (ibid.).

most clearly visible in the continuous 'brain circulation' generated by immigrant entrepreneurs (Saxenian, 2001). Thus, both 'classic' positions appear to be insufficient in that they tend to undervalue this dialectic and, above all, pay too little attention to the changes in the relation between innovative actors and their environment over time.

Ambivalences of New Regionalism The diverse advocates of new regionalism did not only add to the perception of regional studies in the social science community; they also contributed to the further development of social sciences as a whole challenging traditional views of the economy by putting economic action and organization into its wider social and spatial context and by getting concepts like 'social capital' and 'trust' on the agenda even of mainstream economic thinking. Nevertheless they have been subject to serious criticisms, from several directions and academic standpoints, above all since they proved unable to show how the contribution of spatial concentration and social embeddedness to economic competitiveness is in fact as well as how the spatial and the social interact in this context. As Powell and Smith-Doerr put it, referring to the crucial question of trust generation:

> Proximity, as is found in north-central Italy or Silicon Valley, seems to be both too strong and too weak an explanation for trust. Too strong in that the apparent advantages of the industrial districts seem insurmountable: How could models of production that are not as spatially concentrated generate comparable levels of trust? But too weak in that other regions that combine similar skills and advantages cannot reproduce comparable norms of reciprocity and civic engagement. The simple fact of proximity among companies reveals little about their mode of organizing. The vibrancy of the districts is not due to their geography alone, but to their social practices. What other kinds of social arrangements, then are likely to generate trust? (Powell and Smith-Doerr, 1994, p.387).

This is all the more significant in the light of the argument that, as shown in the last paragraphs, a too strong embeddedness – and thus actually a too high level of trust – can also turn into a barrier to change and innovation. As Pierre Veltz (1997) summarizes the discussion new regionalism did not, respectively still does not provide any clue for understanding the interaction of 'spatial and social proximity', except suggesting a congruence, explicable, if at all, either as being the case in a restricted set of regions in a limited period of time or as an idea that the success of regional production systems is supported by the persistence of a traditional identity. This also may have been the case in some specific success regions; yet it does not serve as a general model of how the economy is embedded in its social and spatial environment, particularly given the continuously increasing scale and interconnectedness of economic activity in the globalized world we live in (Amin and Robins, 1991).

New Centrality: Nodes of Knowledge and Power

It is this enormous leap in scale and connectedness which a second key example of how the economy interacts with the social and spatial context it is embedded in is based on, this issue however being developed largely separately from the discussions around the new regionalism. The main focus and starting point was the spatial impact of globalization and information and communication technologies. Unlike assumed by lots of analysts the technical potential to relocate any imaginable economic activity to any place in the world, supported by an increasing immateriality of production, and, thus, the ability to avoid the congestions of urban space, has not led to the annihilation of cities. On the contrary, precisely the cutting edge activities of the global economy tended to be located in a handful of urban master hubs. Saskia Sassen, as the key author behind this 'global city' concept (Sassen, 1991), argues that technology does allow a dispersal of economic activities thereby increasing the 'scale' of action; yet this shift away from previous centres of economic activities has to be understood as an encompassing process of globalization equally adding to the 'complexity' of economic activity (Sassen, 1994).

Andrew Leyshon, drawing on the arguments of Doreen Massey (1995) nicely summarizes the processes under way based on the 'space shrinking' power of communication technologies: They bring about, respectively foster both a deepening and an extension of the division of labour which in turn implies a dispersal of economic activities, at the same time however enabling economic actors to exploit spatial differences much more consciously (Leyshon, 1995). This widening of options to decide between different locations is particularly evident in the case of the global financial system. Money is thus, on the one hand, a means to 'time-space distanciation' (Giddens, 1990). On the other hand, global finance paradigmatically tends to deliberately invest in promising places and to exclude others.

The mechanisms of exclusion also constitute a key to the understanding of Manuel Castells' 'space of flows' in which the key to economic success is the question of having or not having access to global flows. This basic structural logic is, on the one hand, indeed placeless, but it has an important material dimension consisting of 'at least three layers of material support' (Castells, 1996, p.412), through which the logic of place and the patterns of spatial inequality are reproduced: the technical infrastructure, the spatial organization of its managerial elites and the hubs and nodes which act as places of exchange and coordination or as places for 'the location of strategically important functions', respectively.

Thus there is a strong case for the persistence – and even for the increasing importance – of urban nodes in a globalized world, based on the fact that economies of urbanization can be consciously exploited in a world-wide reaching spatial division of labour and that thereby they materially underpin global economic activities. Yet the new centrality is not just a matter of 'material support', and even Castells' layers do not constitute mere infrastructural anchors of a global space of flows. There are two aspects signalling that today's urbanization economies have an important social dimension, on the one hand based on the 'knowledge' necessary to manage the increasing 'scale and complexity' of economic activities (Sassen, 1994), and, on the other, visible in the power relations within the global flows which are manifested both in the structures of the contemporary urban society and in the materiality of urban space (Castells, 1996).

Knowledge: The Social as an Outcome The key to the understanding of Saskia Sassen's idea of a new centrality are the 'finance and specialized service firms', on the one hand necessarily emerged due to the enormous qualitative leap in the organization of the overall economy based on the 'internationalization and expansion of the financial industry' as well as the growing service intensity of economic activities as a whole, on the other hand, tending to be concentrated in 'global cities' 'as sites of production, including the production of innovations, in these leading industries; and [...] as markets for the products and innovations produced' (Sassen, 1991, p.3f.).[5]

The reason for this concentration alleged by the author arises out of the locational features of these specialized services, respectively of producer services in general:

> Producer services, unlike other types of services, are mostly not as dependent on vicinity to the buyers as consumer services. Hence, the concentration of production in suitable locations and exports, both domestically and abroad, are feasible. Production of these services benefits from proximity to other services, particularly when there is a wide array of specialized firms. Such firms obtain agglomeration economies when they locate close to others that are sellers of key inputs or are necessary for joint production of certain service offerings. This would help explain why while New York City continued to lose headquarters throughout the decade, the number and employment of firms servicing such headquarters kept growing rapidly (ibid., p.104).

That the agglomeration economies of specialized service firms are strongly socially underpinned is however only implicitly mentioned by Sassen – in that 'the strategic role of these the specialized services as inputs raises the number and value of top-level professionals' (Sassen, 2001 p.83). The new centrality of complex economic activities thus is actually the spatial manifestation of Daniel Bell's knowledge-class of professionals. However, whenever Sassen uses Bell's knowledge-class as prototype of the urban professional, this occurs in direct relation with the social polarization linked to it. Even the identification of 'talent' as key input for specialized services is used as an argument for showing the downward cycle people unable to cope with the enhanced demands on talent get caught in. The professionals are not explicitly conceptualized with regard to their knowledge input into a complex global economy but merely as an active force in producing inequality on the urban labour market. Sassen's key argument is thus a

---

5   See also the work of the Global and World City (GaWC)-network which attempts to map the global pattern of specialized services (Taylor et al., 2002).

critique of Daniel Bell's optimism regarding the quality of employment in the post-industrial economy. In her words:

> High-level business services [...] are not usually analyzed in terms of their production process. Thus insufficient attention has gone to the actual array of jobs from high-paying to low-paying that are involved in the production of these services. In fact, the elaboration of a financial instrument, for example, requires inputs from law, accounting, advertising, and other specialized services. [...] The production process itself, moreover, includes a variety of workers and firms not usually thought of as part of the information economy – notably, secretaries, maintenance workers, and cleaners. These latter jobs are also key components of the service economy. Thus, no matter how high the place a city occupies in the new transnational hierarchies, it will have a significant share of low-wage jobs thought of as somewhat irrelevant in an advanced information economy, even though they are an integral component (Sassen, 1994, p.105).

Thus even though conceiving of cities as 'post-industrial production sites', which offer the appropriate environment as 'an extremely intense and dense information loop' (Sassen, 2001) for specialized service firms to underpin their non-standardized, non-routine global operations, and even though identifying that the importance of this urban environment is based on a 'combination of firms, talents and expertise' (ibid.), thus naturally having to be anchored in the urban society, the social dimension of these sites is merely dealt with either as the professionals' demands in terms of 'amenities and lifestyles large urban centres can offer' (Sassen, 1995) or as an outcome, an unequal social structure produced by the dominant forces of and each city's role in an increasingly complex global economy. Michael Storper, in a comprehensive critique of Sassen's work, stresses that she, along with her followers, neglects the active role of the urban society: 'Ironically, these theories of global and dual cities seem to exclude what is specifically urban about the process of social polarization – the way that specific practices not only transmit, but transform these global forces into concrete social patterns and relations' (1997, p.233).

Power: The Social is Global A more 'active' role is attributed to 'social patterns and relations' in Castells' concept of a space of flows. Although equally focusing on a polarized urban society the author does not derive this from a dual structure of labour inputs into the service production process but from the social pattern of the dominant global flows. Given that social status is a matter of having or not access to these flows, the key mechanism of polarization is *exclusion*. Such as the position of hubs and nodes in the global hierarchy depends on whether and how they are linked to dominant flows, the same is evident for persons, social groups etc. (Castells, 1996). This exclusion in turn is socially constructed. The space of flows is thus

> ... the spatial logic of dominant interests/functions in our society. But such domination is not purely structural. It is enacted, indeed conceived, decided and implemented by social actors. Thus, the technocratic-financial-managerial elite that occupy the leading

positions in our societies will also have specific spatial requirements regarding the material/spatial support of their interests and practices. [...] Articulation of the elites, segmentation and disorganization of the masses seem to be the twin mechanisms of social domination in our societies. Space plays a fundamental role in this mechanism. In short: elites are cosmopolitan, people are local. The space of power and wealth is projected throughout the world, while people's life and experience is rooted in places, in their culture, in their history (ibid., p.415f.).

Thus social inequality is a matter of uneven distribution of power between, respectively also within networks, implying dynamics of both self-reinforcement and path-dependency, via unequal access to resources and via 'linkages among (different) elite groups' which 'create a cohesive power elite' (Powell and Smith-Doerr, 1994). The manifestation of social inequality in space, which Castells points out, fits into those elite strategies of power consolidation, materialized in the segmentation of urban space through the mechanisms of real estate markets, but also in cultural distinction through cosmopolitan lifestyles, architectural symbolic etc.

Nevertheless, even though Castells considers the social patterns as active elements of a global space of flows, he does not, or even less than Saskia Sassen, conceive of them as local, or as 'specifically urban' patterns and practices. The local dimension of the urban node is reduced to a kind of symbolic footprint of global forces or, to the people without access to dominant flows. Taylor et al. (2002) very nicely describe the differences between Sassen and Castells regarding the relation between cities and the global flow (or network) economy: In Sassen's perspective, 'cities "run" the networks' – although the author does not really elaborate how they do that – whereas, according to Castells, 'the networks "generate" cities'.

Thus, the idea that the social dimension of a particular spatial pattern consists in the power particular places have through their role within global (social) networks tends to overlook the specific local mechanisms behind this power, or to ignore the question 'what makes (particular) cities powerful?' (Allen, 1999) and others not, except deriving it from an external logic of unequal power distribution in a global network space. Allen points out that 'power is produced', and that there are three aspects behind that: first, people's 'social interaction', second, the 'mix of resources' this interaction relies on and, third, the 'practices of power', that is, the actual patterns of power exertion (ibid.). Castells does not provide any hint as to how particular local relations, resources and practices make some cities more powerful than others. Instead he suggests an understanding of cities as mere modes of 'socialization' of the 'new (global) information elite' (Storper, 1997), that is, he considers the social structure and social relations in cities equally driven by external – however more social – forces as Sassen does it.

Ambivalences of New Centrality Both Sassen and Castells without any doubt have the merit of having put 'the urban' back into the centre of academic interest in a globalized era. However, also they have been subject to strong criticisms, above all for the one-dimensionality of defining cities merely through their role in the global economy, thus conceiving of them as mere '*machine(s)* [...], as subassemblies in the overall mechanical structure of the forces and flows of the global capitalist society' (Storper, 1997, *original emphasis*). Thereby, with 'complexity and knowledge' and 'power', they also offer the concepts to grasp the social foundation of the agglomeration economies, yet seeing also the social dimension either as part (in the case of the space of flows) or as a mere (negative) outcome of this machine-like behaviour of capitalism. Their arguments can thus be accused either of ignoring the main lesson of the new regionalism, that is, to see the economy essentially as being embedded in social contexts, or – if assuming that Daniel Bell's knowledge class constitutes the social base of the global knowledge economy – of conceptualizing the social in a too straightforward manner of a uniform and global class logic accomplishing a specific role in societal development. If thus new regionalism oversimplified the interaction between the social and the spatial dimension of economic development by assuming a mere congruence, the new centrality basically made the opposite assumption, suggesting a complete disarticulation between the social foundation of economic activities and the place in which this foundation is 'at work'.

Out of the critique of this kind of oversimplifying the urban some work has arisen which stresses the character of urban nodes as 'places', without neglecting the openness and institutional diversity characteristic for cities (Moulaert and Lambooy, 1996), focusing on the increasing 'variety' and 'pace' of economic action and its underpinning through the diverse conventions and relations of an urban society (Storper, 1997), stressing the 'openness to future development' characteristic for a diverse 'economy *of* the city' (Läpple, 1998)[6] or seeing specific professional communities as 'milieus in the city' which, in terms of their behaviour, equal the innovative regional high-tech milieus of the new regionalism debates without comprising, or being congruent with the whole region, respectively the whole city (Camagni, 1999).

Given the exemplary character of the discussed key debates on an increasing interwovenness of economic organization, social relations and spatial structures in this section we do not provide an encompassing review of this literature; instead we shall now focus on a third thread of debate, closely linked to the work both on new regionalism and on new centrality, but taking us closer to the core of this book: the rise of culturally informed (creative) activities to becoming part of the mainstream capitalist economy.

---

6   Läpple proposes an understanding of an 'economy *of* the city', based on 'untraded mechanisms' and on this openness, opposed to the classic economist concept of an 'economy *in* the city' in which the 'classic' rules of the economy are applied to urban space (Läpple, 1998).

The Creative Economy: Space and the Discovery of the Individual

Culturally informed industries and activities have nearly always constituted important case studies since the beginning of the debates on both a new regionalism and a new centrality. Paradigmatically, the reorganization of the Hollywood movie cluster after the rise of private television and the famous antitrust Paramount decision was described as a role model of flexible specialization underpinned by localized patterns of social relations and the institutional infrastructures of the Los Angeles region (Storper and Christopherson, 1987; Christopherson and Storper, 1986). In addition, albeit not made explicit, the Hollywood case reaches beyond the mere congruence of spatial proximity and social inclusion given that the cluster is only one part of a large and diverse metropolitan economy (Scott, 1996, 1997).

Nevertheless, also the motion picture variant of the new regionalism provoked serious criticisms. The main allegation was that a flexible specialization perspective too strongly stressed small players and the regional level as the essential elements of the reorganization process. As Aksoy and Robins (1992), the protagonists of critique, argued 'the restructuring of Hollywood is being shaped and driven by forces that are not adequately recognized or emphasized in the "flexible specialization" thesis' (p.6). In their view it was not primarily the reorganization of production that mattered but the access to and the extension of markets. As a result, they held, 'the emerging industrial structure is made up of a very small number of very large companies which dominate the market through their position as gatekeepers to the distribution of filmed entertainment on a global scale' (p.11).

The sharpness of positions in the debates on localizing and globalizing forces in the prime sectors of the creative industries has been mitigated in the meantime, at least to a certain extent. More recent accounts consider organizational concentration on the distribution and decentralization on the production side as complementary rather than opposite trends in the reorganization of the industry (Lash and Urry, 1994; Scott, 1997). On top of that, they do not only focus on the internal logic of the industry but consider the overall creative industries as playing a forerunner role in a wider socio-economic transformation, thereby signalling a more general change as to the relation between economy and culture. Regarding our focus on the interaction of economic organization with its social and spatial environment two different approaches to this assumed 'cultural turn' (Crang, 1997) of the economy can be identified: an *output* oriented approach stressing the increased cultural 'content' of the capitalist economy and its implications for the spatial pattern of production and distribution (Scott, 1997), and an *input* perspective orbiting around a changing work force and underlining the fact that this change is oriented along the role model of the classic cultural and artistic professional; this in turn changes the general demands in terms of 'where' work takes place in a significant way (Florida, 2002a).

The 'Cultural Economy of Cities' As pointed out at the outset of this book a seeming convergence between previously separated spheres of economy and culture had already been dealt with in the 'mass culture' literature of the Frankfurt school (Adorno, 1991), yet with a particular viewpoint highlighting the 'expansionary and imperialistic' logic of capitalism and the consequent subordination of culture under 'the cash nexus and the logic of capital circulation' (Harvey, 1989a). This political economic view of a 'commodification of culture' has in the meantime been transformed, on the one hand into a pragmatic perspective of culture industries as an important sector within the process of economic regeneration (Hudson, 1995; Kunzmann, 1995; AG Kulturwirtschaft, 1995). As Allen Scott puts it drawing on empirical work on the wider 'multisectoral image-producing complex' of the Los Angeles region:

> An assumption frequently made by local economic-development practitioners is that high-technology industry represents the one best pathway to regional prosperity. As the present investigation has shown, however, cultural-products industries can also be an extremely powerful vehicle of job creation and growth (Scott, 1996, p.319).

On the other hand, the 'multisectoral' character of Scott's cultural-products industries points to more than the emergence of culture to a promising new 'economic base' of cities (Zukin, 1995). The interrelatedness of culture and economy and its spatial dimension is more complex given the fact that, as Scott holds, 'an ever-widening range of economic activity is concerned with producing and marketing goods and services that are infused in one way or another with broadly aesthetic or semiotic attributes' (1997, p.323). Lash and Urry even accentuate this argument by holding that the logic of a commodification of culture stressed by Horkheimer, Adorno and their followers is actually reversed in the post-industrial (or as they put, 'post-Fordist') economy:

> We are arguing, pace many Marxists, against any notion that culture production is becoming more like commodity production in manufacturing industry. Our claim is that ordinary manufacturing industry is becoming more and more like the production of culture. It is not that commodity production provides the template, and culture follows, but that the culture industries themselves have provided the template (Lash and Urry, 1994, p.123).

It is again Allen Scott who provides the work related to the spatial dimension of this change. In his influential paper 'The Cultural Economy of Cities' he holds that, similarly to how Saskia Sassen sees the spatial organization of specialized services in a handful of global cities, the 'post-Fordist global economic order' will make 'the heartlands of modern world capitalism – places like New York, Los Angeles, London, Paris and Tokyo, [...] – continue [...] to function as *the bulwarks of a new cultural economy of capitalism*' (Scott, 1997, p. 324, *emphasis added*).

This key position of a handful of urban 'master hubs' in a global cultural economy is essentially based on a duality of local production and global distribution characteristic for the industry in question. On the production side, these big cities display a particular symbiosis of 'place, culture and economy' consisting of 'massive urban communities characterized by many different social functions and dense internal relationships' (ibid., p.325). This coexistence of difference and density stimulates 'cultural experimentation and renewal' as major input into the cultural economy and provide the necessary economic strength in order to underpin cultural production. Culture and economy in the cities thus display a virtuous circle of mutual reinforcement. As Scott puts it:

> Local cultures help to shape the nature of intra-urban economic activity; concomitantly, economic activity becomes a dynamic element of the culture-generating and innovative capacities of given places. This comment applies, of course, to forms of economic activity that are concerned with non-cultural as well as cultural products (ibid.).

Yet, the interrelations of place, culture and economy are even more complex, also – indirectly – having an impact on the distribution side. Given that the local cultural production complex is in a 'recursive relation' with the cultural attributes of place, the outputs of the industry contribute to the specific image of the city, in turn being 'assimilated back into the city's fund of cultural assets where they become available as inputs for new rounds of production' (ibid.). That is to say, the local cultural production system is based on a sort of Marshallian district atmosphere stimulating the creativity of its members, which yet is constantly fed by the reflections of its output into the system itself. At the same time, of course, the amalgamation of production system and local image function as marketing tool for the very output. The distribution side itself is – as pointed out by Aksoy and Robins for the Hollywood motion picture cluster – constituted by a small group of large 'gatekeepers' to the global markets which are however also part of the local cluster thereby benefiting from both the production system and the image attached to its symbiosis with the place.

Scott's arguments, further developed by himself for particular local creative clusters (Scott, 2000), and by others, regarding the changes in the local 'production landscape' (Hutton, 2000), the global organization of single cultural products industries (Krätke, 2002) and the side-effects of a world-wide distribution in the local production systems of different markets through the necessity of embedding messages there (Brito Henriques and Thiel, 2000), do not only stress the major importance of big cities in culture informed industries catering globally; they also provide clues as to the interrelation of the economy's organization with the urban society, revoking classic accounts on complex and diverse urban environments as incubators of cultural experimentation and innovation.

However, the intricate symbiosis between local production system and the image linked to its attachment to a certain place, while being certainly true for the location of the prime sectors of a creative economy in the corresponding 'flagship' cities, epitomized by Hollywood films, Parisien haute couture etc., does not underpin a general trend towards a new 'template' for capitalist production provided by the cultural industry. It is rather Scott's fruitful point regarding the diversity of urban society and its stimulating effect for cultural production which helps us in this context, taking us to the input side of an increasingly creative economy, and to Richard Florida who thinks about the general changes on this side as the 'sea change' Western industrial societies are currently experiencing.

Cities and the 'Creative Class' Florida's 'The Rise of the Creative Class' was certainly one of the most prominent and successful academic books not only within the regional science literature of the last years. His main point is that the forerunner role which Lash and Urry, among others, attribute to the culture industries as to the restructuring of the overall economy is primarily a social phenomenon, thus chiefly concerning the work force side. Given that economic success and competitiveness is increasingly based on new ideas, and consequently is more and more 'powered by human creativity' (Florida, 2002a), patterns of work and life previously restricted to a relatively small group of privileged creative professions have now diffused into the core of capitalist work. As he puts in more detail:

> Artists, musicians, professors and scientists have always set their own hours, dressed in relaxed and casual clothes and worked in stimulating environments. They could never be forced to work, yet they never were truly not at work. With the rise of the Creative Class, this way of working has moved from the margins to the economic mainstream. While the no-collar workplace certainly appears more casual than the old, it replaces traditional hierarchical systems of control with new forms of self-management, peer recognition and pressure and intrinsic forms of motivation, [...] (ibid., p.13).

On the macro-level of the society this change concerning the input into production triggers a broader shift in the social structure. That is to say, it both is based on and produces a 'Creative Class' of a non-conformist work-force previously not having been integral part of the capitalist economy. Again in Florida's own words:

> The increasing importance of creativity, innovation and knowledge to the economy opens up the social space where more *eccentric, alternative* and *bohemian* types of people can be integrated into core economic institutions. Capitalism, or more accurately new forms of capitalist enterprise [...] are in effect extending their reach in ways that integrate formerly *marginalized* social groups and individuals into the value creation process (Florida, 2002b, p. 57, *emphasis added*).

This general social dimension of change in turn, and in our perspective most importantly, has an important spatial component implying that economic development takes place – roughly put – where the members of this creative class like to be. Also, this impinges on the corresponding policy strategies:

> Building a vibrant technology-based region requires more than just investing in R&D, supporting entrepreneurship, and generating venture capital. It requires *creating lifestyle options that attract talented people*, and *supporting diversity and low entry barriers to human capital*. These attributes make a city a place where talented people from varied

backgrounds want to live and are able to pursue the life they desire. (Florida and Gates, 2001, p.7, *emphasis added*).

Florida's key concept regarding the competitiveness of places is, in analogy with – but simultaneously in disassociation from – the concepts of 'human capital' put forward by Glaeser (1998) and 'social capital' stressed by Putnam (2000) and others, the existence of 'creative capital', that is, the presence of a particular type of 'non-conformist' people and the 'underlying factors that shape the location decisions of these people' (Florida, 2003). These in turn consist of an open and tolerant cultural climate in which the non-conformist people are accepted and a rich and diverse 'cultural infrastructure' accomplishing their high demands regarding lifestyle.

The idea of the 'Creative Class', besides being presented in a well written and stimulating manner, contributes in a significant and consistent way to the understanding of how economic activity is interrelated with its socio-spatial environment. Particularly through the focus on the input side of the economy both the social rootedness and the spatial variation of economic development are conceptualized in a very innovative and convincing way. The arguments strongly resemble Daniel Bell's idea of a class of professionals as social expression of the post-industrial society (Bell, 1974). Yet, in addition to Bell, Florida stresses the individualistic nature of the emerging new class, thereby not only describing a shift in the content and nature of economic activities but also underlining the fact that this shift poses the need for the economy to adapt to the work force in terms of work organization and work environment. Moreover, also the spatial dimension of the rise of the creative class is derived from this necessity. It seems no accident that, when talking about work organization, Florida confronts William H. Whyte's 'Organization Man' as archetypical description of work in the industrial society with Jane Jacobs' 'Death and Life of Great American Cities' as providing the role model for organization in the new 'creative' era (Florida, 2002a).

Nonetheless, the fact of having chosen Jane Jacobs as advocate of a context in which creativity is stimulated also shows the shortcomings of the arguments. Explicitly confronting 'organization' with 'creativity' and replacing thus the work process of the industrial era with the mere idea of a stimulating environment[7] fails to conceive of the ways how industries thriving on the creativity of individual workers really work. Besides a series of catchwords like 'the no-collar workplace', 'soft control' or the polarity of 'the white-collar sweatshop vs. the caring company' the inside of an economy based on the creative class remains rather obscure (Thompson and Warhurst, 2004). Thus, while Florida argues conclusively about the new opportunities and challenges brought about by a non-conformist workforce, he does not suggest any clue for understanding how economic actors deal with them, except providing 'talents' with an inspiring and tolerant

---

7   By entitling the section about Whyte's and Jacobs' arguments with 'creativity vs. organization' Florida suggests rather the idea of a post-organization era than of a new model of organization (Florida, 2002a, p.40).

environment. Jane Jacobs' diverse urban neighbourhoods so to speak epitomize such an environment; yet the argument, being underpinned by a thorough statistical analysis, too quickly suggests a linear relation between creative people, the environment in which they like to live and the economic output they produce. Thus similarly to Daniel Bell and his followers also Florida – despite clearly emphasizing individualism and thus differentiation as the dominant tendency in economy and society – fails to conceive of how the dominant forces are not just 'transmitted' into the functioning of the economy and its environment but 'transformed' 'into concrete social patterns and relations' in space (Storper, 1997).

The Ambivalences of the Creative Economy Approach The idea of a growing convergence between the spheres of culture and the economy has strongly influenced the overall human and social sciences, not only regarding the relation between both spheres in socio-economic reality, but also in methodological terms, here above all strengthening perspectives seeing socio-economic phenomena through a cultural lens (Crang, 1997). The extreme body of work risen in this context has also, and similarly to the debates put forward in the preceding sections, contributed to the increasing perception of space as an important category regarding the functioning of a creative economy, and the two perspectives just outlined constitute good examples for that, seeing the spatial environment both consisting of a manifestation of the social substratum of economic activities and impinging on this substratum through its built materiality and the images linked to it. Yet the cultural lens has also tended to obscure the material nature of economy and society to a certain extent; or in Sayer's (1997) words, 'the flip-side of this has been a decline in interest in political economy', and the two approaches to a culturalization of either products and services or work patterns do also reflect this side. In this context, two aspects symbolize their shortcomings in terms of the 'practical' and procedural character of a converging industry both regarding its substance and its social foundation through a specific type of personality as worker.

First, the assumed convergence can neither be understood as a harmonious merger of both spheres nor as a subordination of culture to the economy or vice-versa. An approximation of both spheres as suggested by Lash and Urry, Scott and others implies that also the different basic logics behind culture and the economy come closer together and necessarily interact. Andrew Sayer, labelling this interaction of differences as the 'dialectic of culture and economy' (ibid.) points out that both spheres are in fact 'distinguishable' in that culture is basically 'non-instrumental' whereas the economy 'involve(s) a primarily instrumental orientation'. 'Cultural norms and values [...] are seen as good and bad in themselves rather than in terms of their consequences' (ibid., p.17). Even though one can criticize this perspective as conveying a too positive view of culture[8] it is certainly true that the 'cultural turn' involves an increasing tension between

---

8   Sayer (ibid.) himself raises this point, drawing on Bourdieu's concept of 'distinction' and the role of 'cultural capital' in this context.

culturally and economically shaped 'frameworks of action' (Storper, 1997) which has to be dealt with in the very process of cultural production and commercialization. The evidence presented in the advertising case study in chapters 3 to 5 will show this in a very drastic way.

Second, particularly in Florida's arguments there is a conceptual contradiction between the focus on an individualist work force, on the one hand, and the maintenance of a homogenizing class concept on the macro level of society, on the other. The quasi mechanic relation between creative people and the economic success of certain places is basically a direct consequence of this homogenization. If one takes Florida's ideas of individualism and non-conformism seriously, one has to stress precisely the dissolution of traditional social structures of strata, classes etc. (Beck, 1991) as dominant trend which in turn also refracts this linear relation given that in an individualized society human action is decreasingly self-explaining, respectively ever less relying on a pre-given and thus self-evident set of behaviour. In the words of Giddens (1990) and Lash and Urry (1994), the essence of contemporary societies consists in an increase of 'reflexivity' regarding the way subjects act.

In sum, the cultural economy approaches too strongly emphasize the role of the social and the spatial as positive stimulation for cultural production, that is, creative actors are embedded in specific locally concentrated communities which additionally benefit from the aesthetic and infrastructural environment they are surrounded by. While this opportunity-oriented perspective certainly contains a lot of truth about a creative economy, it only tells half of the story. The remainder of this chapter will therefore elaborate on another dimension of the supposed 'cultural turn', focusing on the risks, frictions, ambivalences and complexities inherent to it which have to be solved on the operational level of the work and life of 'reflexive' individual actors.

Recontextualization through the Individual: Reflexivity and the Convergence of Work and Life

The term 'reflexivity' as a concept grasping the ongoing changes in contemporary societies had its origin in a debate on 'reflexive modernization' (Beck et al., 1994) having started in the middle of the 1980s and suggesting a counter-concept to the fashionable mainstream social science thinking which conceived of the current era of capitalism either as the end of something else (visible in adding a prefix post- to the previous labels: post-industrial, post-modern, post-Fordist) or as a change from one 'primary logic' of social organization to another (knowledge, information etc.). Unlike this, the 'reflexive modernization' approaches characterized the ongoing changes as a consequent implementation of the principles of modernity; according to Giddens (1990) we are rather experiencing a period of 'radicalized modernity' than a post-modern era.

'Reflexivity' in this sense refers to the fact that 'all human beings (are) in touch with the grounds of what they do'; all kinds of human life therefore include 'a reflexive monitoring of action' (ibid., p.36). In traditional societies this monitoring takes place in the light of established rules of the local societies which in turn are based on the past experience of previous generations. In Giddens' words:

> Tradition is a mode of integrating the reflexive monitoring of action with the time-space organization of the community. It is a means of handling time and space, which inserts any particular activity or experience within the continuity of past, present and future, these in turn being structured by recurrent social practices. Tradition is not wholly static, because it has to be reinvented by each new generation as it takes over its cultural inheritance from those preceding it. [...] However, in pre-modern civilizations reflexivity is still largely limited to the reinterpretation and clarification of tradition, such that in the scales of time the side of the 'past' is much more heavily weighed down than that of the 'future' (ibid., p.37f.).

In modern societies this weight of the past is no longer valid without restrictions. This is not to say, 'reinterpretation and clarification of tradition' does not occur any more; however, traditional practices are, as all social practices, 'constantly examined and reformed in the light of incoming information about those very practices, thus constitutively altering their character' (ibid.). As a result action is no longer self-evident or pre-given by the action of previous generations but has to be re-considered according to possible other actions etc. Thus, individual agency is 'set free' from traditional structures, thereby being confronted with an enormously enhanced variety of options which in turn implies an increasing uncertainty about which is the right choice. An increase in reflexivity thus basically means that 'reflexive monitoring of action' in modernity inherently involves 'radical doubt' (Giddens, 1991) thereby being a disproportionately more complex and challenging process than in a pre-modern context. Giddens holds that the increase in reflexivity affects all spheres of human life; in his book 'Modernity and Self-Identity' he illustrates it most explicitly with regard to the way modern individuals have to build their own biographies (ibid.). In radicalized modernity, he argues, 'the self [...] becomes a "reflexive project"' which 'consists in the sustaining of coherent, yet continuously revised, biographical narratives' (ibid. p.5).

This increase in reflexivity on the level of the individual also includes the economic part of the biography, as the development of a professional career etc. In addition, according to Lash and Urry, it also converges with a functional change within the economic realm itself, in that 'post-Fordist production' is getting increasingly 'reflexive' requiring 'employees as agents' to take 'more individual responsibility' (Lash and Urry, 1994, p.122). The authors continue:

> This sort of economic actor is no longer to such a great extent circumscribed by the constraints of 'structure', subject to the rules and resources of the shopfloor. Instead he/she operates at some distance from these rules and resources; he/she makes decisions as to alternative rules and resources; and he/she finally is responsible for the continuous transformation of both shopfloor rules and (in process and product) resources (ibid.).

Thus, also in their life as economic actors, modern individuals are 'reflexive subjects' their agency being 'set-free from heteronomous control or monitoring' (ibid., p.4). Interestingly, however, this individualization is not just a process of emancipation of individual agency from structure but 'it is structural change itself in modernization that so to speak forces agency to take on powers that heretofore lay in social structure themselves' (ibid.).

As a result of these challenges individuals are confronted with both in their lives and at work, the logics of the spheres of work/production and life/reproduction tend to come closer together, due to two reasons: First, given that the professional career is also part of a 'reflexive project of the self' (Giddens, 1991) work becomes involved in the attempt to build a consistent biographical narrative. That is to say, also the development of a professional identity is subject to the ambivalence of an enhanced variety of options, on the one hand, and the enhanced possibility of wrong choices, on the other. Second, work and life both obey to this similar logic of a decreasingly self-evident 'action framework' (Storper, 1997), thus requiring similar attitudes, abilities etc. Radically argued one could even claim a dissolution of Habermas' distinction between systems world and life world, given that the reflexive biography requires elements of 'strategic rationality'[9] as well as reflexive work involves a discursive and communicative dealing with the present work context (see also Baethge, 2001; Grabher, 2004).

Two important criticisms can be raised in this context: First, stressing the convergence of work and life on the level of the individual biography too strongly focuses on individual agency thereby running the risk of building again an 'atomized' or 'undersocialized' conception of the economic actor, even if equipped with a wider set of motives than the *homo oeconomicus*. Michael Piore, in a chapter on work experience in the Italian industrial districts provides an idea of how individualization does not equal atomization. Drawing on Hannah Arendt's distinction between labour, work and action as three basic realms of human activity he attributes the productive activity in flexible specialization as action in the sense that 'the production process becomes for the people who participate within it a public space like the political forum of ancient Greece, that they see that space as realm in which they reveal themselves to one another as individuals' (Piore, 1990, p.66). Following this argument, the economic action of productive 'workers' in the Italian districts provides an important contribution to give 'meaning to people's lives' (p.72). However, the revelation of oneself as individual can only occur within 'a community of equals'. 'If the other members of the community are not like one, they cannot appreciate one's differences' (p.67). That is to say, a positive integration of professional life into a consistent biographical narrative can only come about if it happens in the light of a community able to assess one's professional value.

In addition, indeed reflexive subjects cannot be considered as isolated actors given that the process of modernization does not only occur on the level of the

---

9   Giddens (1991) even talks about 'strategic *life-planning*' as 'substantial content of the reflexively organized trajectory of the self' (p.85, *original emphasis*).

individual, nor does reflexive action replace pervasive economic structures. Giddens explicitly holds that the reflexivity on the level of the biography interacts and conflicts with the disembedding, respectively globalizing forces of 'abstract systems' through which the enhanced variety of options is filtered in various ways (Giddens, 1991). In a summarizing argument he puts this dialectic logic of modernity as follows:

> One of the distinctive features of modernity, in fact, is an increasing interconnection between the two 'extremes' of extensionality and intentionality: globalizing influences on the one hand and personal dispositions on the other (ibid., p.1).

These abstract systems, consisting of so-called 'expert systems' and 'symbolic tokens' the most important of which is money, are the main driving forces of the 'time-space distanciation' or 'disembedding', that is, the '"lifting out" of social relations from local contexts of interaction' and their stretching over space (Giddens, 1990, p.21). Yet, also the increase of 'faceless commitments', as Giddens terms the disembedded interpersonal relations which do not require 'circumstances of co-presence', 'interact(s) with re-embedded contexts of action which may act either to support or to undermine them' (p.80). That is to say, 'faceless commitments' necessarily rely on 'facework commitments'; his argument in this context is that relations 'stretched' over space basically need *trust* in the disembedding mechanisms, that is, in expert systems and symbolic tokens. The building of this trust does not occur through the same mechanisms as in traditional societies in which past personal experiences and norms and values existing in the collective memory of localized social systems quasi automatically function as trust generators. Thus, under conditions of modernity trust, of both lay actors in abstract systems and of different members within 'abstract' expert systems has to be actively built. Between lay actors and expert systems this occurs through personal 'access points' to 'representatives of expert systems' in order to make oneself sure of their integrity (p.88).

Within expert systems – in which we are interested more strongly given our focus on individuals' professional biographies – re-embedding occurs through facework relations between experts which, according to Giddens, have two functions: first, to 'sustain collegial trustworthiness' and, second, to provide professional actors with tools to cope with the 'reflexively mobile nature' of their own activity (p.87). Put in another way, abstract expert systems which on the one hand are argued to be the main motors of the disembedding of social relations, on the other hand, in their very functioning, indispensably need facework rituals of personal relations in which both interpersonal trust and a common understanding of a complex and uncertain activity is generated. Thus, unlike lots of commentators hold in relation with the social foundation of a knowledge economy, it is less the need for learning and quick adaptation to changing demands which fosters the embeddedness of economic systems in social relations but rather the need of a de-traditionalized economic activity to generate both the professional identity and the social capital no longer provided in a self-evident way.

The second criticism is less theoretically informed, addressing the assessment of the ongoing changes. Simply put it considers the idea of a convergence of the work and life of reflexive individuals as too optimistic. There is a whole body of political economic accounts focusing on the increasing precariousness of work patterns and labour relations as the dominant tendency of a post-industrial economy (Carnoy, 2000; Beck, 2000, among others). The contribution most strongly related to the 'reflexivity' debate, however without explicitly elaborating on it, is provided by Richard Sennett, in his book on 'the personal consequences of work in the new capitalism', programmatically titled 'The Corrosion of Character' (Sennett, 1998). Drawing on the terms employed above, Sennett's key argument is that, although it is true that the structural forces of contemporary capitalism require a more autonomous action, thereby to a certain extent freeing individual agency from structure, this implies a compulsion to flexibility and adaptability in terms of the personal career which obstructs rather than supports the construction of a consistent biographical narrative. The consequences of this are, at least on the level of highly skilled post-industrial professionals, less a matter of material precariousness, but rather affect the personality and identity of individuals in terms of their personal well-being.

Even though Sennett's contention resembles the 'global-capitalism-as-a-machine'-accounts pointed out above thereby presenting an all-too structuralist idea of the current functioning of the economy and focusing rather on its impact on than on its interaction with society, it touches certainly on one important point: that the outlined convergence of work and life inherent to an overall increase in reflexivity does not mean a harmonious merger of both spheres. The complex 'interconnection' between 'globalizing influences' and 'personal dispositions' put forward by Giddens of course also influences the relation between work and life. Also Lash and Urry's argument of reflexivity within the economic sphere does not imply a pure and agreeable freeing of agency driven by structure. Even if it is structure that needs agency independent from it, there is still a fundamental contradiction between structure's interest of making use of the individual's capabilities (within the context of a global 'flow economy' (Lash and Urry, 1994)) and the individual's interest of acting independently (both within the work context and in the context of his or her biography). These contradictions inherent to the convergence of work and life have to be carried out, and, logically, the 'arena' in which this occurs is the labour market of precisely the reflexive economic activities.

Labour Markets as 'Arenas' of a Conflicting Convergence

The labour market is definitely the key institution through which the relation between economy and society is 'put into practice'. This is not only due to its macro-economic role as the key institution for the distribution of the national income. More importantly even, it mediates between the instrumental and rational forces of the economy and the embeddedness of economic actors in social contexts in a paradigmatic way, not only put forward in the recent discussions on a seeming reorientation of the economy. Also historically, it plays a major role for the understanding of how economic organization and social relations interact in the functioning of the economy.

The function of the labour market as a 'hinge' between economy and society in a market economy has been elaborated most clearly in the work of Karl Polanyi (2001 (1944)). Polanyi's key to the understanding of labour markets is the concept of 'fictitious commodities' not only concerning labour but also land and money. As he explains:

> The crucial point is this: labour, land, and money are essential elements of industry; they also must be organized in markets; in fact, these markets form an absolutely vital part of the economic system. But labour, land, and money are obviously *not* commodities; the postulate that anything that is bought and sold must have been produced for sale is emphatically untrue in regard to them. In other words, according to the empirical definition of commodities they are not commodities (ibid., p.75, *original emphasis*).

One of these three fictitious commodities 'stands out': labour, given that its commodification involved that 'human society had become an accessory of the economic system' (p.79). The creation of the 'fictitious commodities' – according to Polanyi – was the crucial precondition for the establishment of a market economy in the 19th century. At the same time, however, the very fact of these commodities continuing to be a mere fiction implies that a market economy is necessarily socially underpinned and – as put at the outset of this chapter – that it can never fully unfold without 'annihilating the human and natural substance of society' (p.3).

That is to say, the interaction of economy and society in the labour market does not only stand for itself as characterizing the functioning of the economy; it also epitomizes the dialectic nature of the modernization process advanced by Giddens between the disembedding dynamic of modernity based on 'abstract systems' – in Polanyi's argument the abstract commodifying logic of the market economy – and the re-embedding forces inherent to it which 'pin them (the disembedded social relations) down to local conditions of time and place' (Giddens, 1990, p.79f.) – in Polanyi's theoretical edifice represented by the fact that 'labour is only another name for human activity which goes with life itself' (Polanyi, 2001 (1944), p.75).

Fred Block, in a nice summarizing introduction to the recent re-edition of Polanyi's masterpiece 'The Great Transformation' holds that the argument of the fictitiousness of labour, land and money can be understood on two levels: First, on a moral level given that 'it is simply wrong to treat nature and human beings as objects whose price will be determined entirely by the market' (Block, 2001). Second, and for our purpose more importantly, on an 'analytical' level there is a need to 'manage' the fictitious commodities in order to avoid the annihilation of the social and natural substance of society and thus consequently the functioning of the very roots of the market economy. As a result the commodity fiction necessarily implies state regulation. As Block puts it:

> In short, the role of managing fictitious commodities places the state inside three of the most important markets; it becomes utterly impossible to sustain market liberalism's view that the state is 'outside' the economy (ibid., p.xxvi).

Regarding the labour market Polanyi thereby roughly explains the need for the emergence of a welfare state's basic elements: '[...] the state has to manage shifting demand for employees by providing reliefs in periods of unemployment, by educating and training future workers and by seeking to influence migration flows' (ibid.) given that the previous subsistence oriented mechanisms of compensating crises no longer function.

Of course the current period of restructuring based on an increasing reflexivity on the level of the individual and an assumed convergence of work and life following from this is less driven by these basic material elements of society. The present reflexive era, seen as a process of individualization, on the one hand means – as pointed out above – a consequent implementation of modernity (Giddens, 1990) and thus a continuation of the disembedding forces of the market economy. In accordance with Beck (1991) it can be understood as a new round of disembedding in which the social foundations of the industrial post-war society – consisting above all in a clear social division of labour between male breadwinner and female housewife – tend to be dissolved as a 'side effect' of its own success. On the other hand, argued on a macro-level, this second round of disembedding is only partly a direct consequence of market forces. This is due to the fact that the success of the post-war model of Western societies was not only based upon the enormous productivity increases in the economy but also on precisely the state policy of 'managing the fictitious commodity' of labour force, through guaranteeing mass markets, and through raising the general level of education.[10] The increased reflexivity thus originated from an enhanced material level of possibilities based on a general wealth increase, but also from people's more demanding attitude towards their lives rooted in the overall growth in cultural capital. This attitude – as discussed above – has affected also the professional part of life.

Nevertheless, the commodity fiction related to labour has remained existent. One can even hold that, based on a specific dimension of this fiction, its impact has been accentuated. Given that it is not labour itself that is bought on the market but a fictitious capacity to work, the labour contract contains an inherent uncertainty regarding the outcome of work (Berger and Offe, 1984).[11] This uncertainty has

---

10 For a concise account of the most important elements of the post-war model of Western industrial societies it is worthwhile to study the work produced by the French regulation school. As an introduction and overview see Leborgne and Lipietz (1990).

11 In this context Berger and Offe (1984, p.91f.), drawing on Marx, talk about the 'indetermination gap' inherent to the labour contract which has to be 'closed in the firm by means of managerial control'.

always shaped the organization of industrial work. Taylorism is usually conceived of as the 'classic' reaction to it, splitting the work process into small units in order to take as much responsibility as possible for the outcome of work away from the single worker (Buttler, Gerlach and Liepmann, 1977). The more reflexive the economy is, that is, the more every single economic decision is subject to doubt about what is right, the higher is the uncertainty about the outcome, yet the less appropriate appears a strategy of removing responsibility from the worker, given the extraordinary costs of planning and control in the light of the complexity and volatility of production processes. The economic dimension of reflexivity pointed out by Lash and Urry (1994) can thus be interpreted precisely as a logical method of dealing with an increased uncertainty by shifting the responsibility for the outcome of a fictitious capacity to work back to the individual worker.

In very general terms, thus, the convergence of work and life based on an increasing reflexivity in both work and life of late modern individuals, from the standpoint of Polanyi's concept of 'labour as a fictitious commodity' can be considered as interaction between, on the one hand, the radicalization of the fictitious nature of labour in the light of an increasingly complex economy and, on the other, a side-effect of the welfare state's dealing precisely with this fictitious nature. Clearly, this convergence involves a series of conflicts between 'the powers of the individual' and 'the powers of markets and technologies' (Touraine, 2002), which are filtered through the labour market, yet not only implying a 'colonization' of the individual 'life world' through the structural forces of the 'systems world' (Habermas, 1981) but rather a multiplicity of uncertainties based on the reflexivity of action lying in both worlds as well as in their complex modes of interaction.

It is an important question how the logic of space may intervene in this intricate interplay of structure and agency in the labour market of reflexive activities. In line with Polanyi the spatiality of the labour market can be simply derived from the fact that – according to David Harvey's famous claim – 'unlike other commodities, labour power has to go home every night' (Harvey, 1989b), that thus the 'local' is the place where 'labour power is socially produced and reproduced' (Peck, 2000).

In this context of the inherent 'localness' of labour markets (Peck, 1996) two points of departure can be drawn from the reflexivity debate: First, the 'social production' of labour power is increasingly demanding given a work context in which the responsibility for the results of his or her work is shifted to the worker himself. Coping with the challenges inherent to this shift is likely to exceed the capacities of a classic national education and training policy, and success and/or failure in this context will consequently produce inequality between regions (Peck, 1994). Second, the uncertainties inherent to the intensified interplay between work and life appear to produce what Alfred Marshall had considered as the advantages of labour pooling in 'thick labour markets' (Krugman, 1991; Glaeser, 1998; Florida, 2002) in order to prevent workers from the consequences of 'industry-specific shocks' by offering alternative work options without implying the need to change also the living place. It seems logical that the multiplicity of uncertainties inherent to reflexive labour markets enhances the advantages of thick labour markets also beyond the ups and downs of industry-specific business cycles. The subsequent advertising case study will provide encompassing empirical material related to the spatial dimension of a reflexive labour market, which then will be deepened in the concluding chapter of this book.

The Economic, the Social and the Spatial: Towards a Subject-Oriented Understanding of Recontextualization

This chapter had a double objective: on the one hand, to outline the specific approach of this book and to frame it by placing it in the context of related debates; on the other, to give at least some hints as to how this approach can be derived from the changes Western societies have been going through during the last 30 years. Put in more concrete terms, the chapter was to give an account of an understanding of post-industrial economic activities as being closely intertwined with the social and spatial context they are embedded in. At the same time it should empirically ground this primarily epistemological position by both showing that and explaining why the rise of a post-industrial economy has resulted in a real increase in the interrelatedness of economic action and its socio-spatial environment thereby having brought about a seeming turnaround in the previous process of societal modernization generally said to be characterized by a continuing separation of the spheres of economy and society and an increasing domination of the latter through the former, understood either as a social progress or as a 'colonization' of society through the forces of capitalism.

Starting from this proposition three exemplary debates strongly related to urban and regional studies were presented and discussed with regard to the way they conceive of how the organization of the economy interacts with the social relations and spatial structures which 'surround' it:

1. The *new regionalism*-debate dealing with the success of specialized regional economic clusters based on the social underpinning of either flexible adaptation to changing markets or of technological innovation.
2. The *new centrality*-thesis focusing on the need for the concentration of knowledge and power in an increasingly complex globalized economy.
3. The work on the *rise of the creative economy* either reflecting an ongoing merger of products and services with cultural attributes thereby both implying the production in 'massive urban' contexts and fostering a close linkage between the output and the image of the place it is produced, or being based on a new social class of individualist creative workers which are considered as the keys to future competitiveness, which yet need, respectively like, to live and work in open and stimulating environments.

Roughly put, all three debates provided important contributions to a wider understanding of the economy and to the grasp of the ongoing change in the production systems of the industrialized world and its relation to the socio-spatial pattern of these systems. However, they are actually weak in conceiving of what is the real driving force behind a recontextualization of economic activities precisely in a post-industrial global knowledge economy. Again in very general terms, in all three debates the basic requirements of a new economic logic are outlined and it is discussed how spatial and social circumstances fit into these requirements, respectively how they are affected by them. To give an example: a fragmentation of mass markets leads to the flexibilization of industrial production; this in turn needs to be both spatially concentrated and socially underpinned.

There are two exceptions to this general pattern of the social and the spatial as positively contributing to the functioning of the industry. The first consists in the observation that specific social and spatial settings can also hinder rather than foster innovativeness. Given that the role model for a context unable to change through the rigidity of its socio-spatial setting is the Ruhr region, that is, a paradigmatic example of heavy manufacturing, this argument does not reflect the historical specificity of the change we have been going through within the last three decades. The second exception is at least partly – Richard Florida's 'Creative Class' argument. He at least insists on the fact that it is society itself that drives the change: 'Society is changing in large measure because we want it to' (Florida, 2002a, p.4) is the somewhat emphatic starting point of his reasoning about the growing importance of creativity. Nevertheless then he also revokes this active role of society, returning to discussing the social and spatial needs of 'an economy powered by human creativity'.

Our own argument is, first, against precisely this passive or reactive role of the social and the spatial in the current socio-economic change and, second, places, again in a further development of Florida's ideas, the logic of linking economic activities with social relations and spatial structures on the level of the individual. Drawing on both Giddens' and Lash and Urry's concept of 'reflexive subjects' we hold that the driving force behind the turnaround of social modernization are the individuals in a more complex and 'reflexive' economic and social environment, and that the re-embedding of economic activities is based upon a 'convergence of work and life' driven by the fact that knowledge-intensive work increasingly requires to act independently from pre-given patterns as well as individual biographies no longer rely on traditional role models.

Both changes on the one hand imply an increase in individual independence and emancipation; however they enhance also the uncertainty about how to act in an appropriate way. On the other hand, the changes on the level of the individual interact and conflict with the ongoing disembedding forces of modernization, and this interaction happens above all on the labour market of knowledge-intensive activities. The labour market is thus the pivotal 'arena' in which disembedding and re-embedding dynamics interact, in which they complement or undermine each other, in which thus – in line with Polanyi's thinking – the inherent conflict between labour as a commodity and labour as human being is transformed into social practice, in the light of an economy in which labour power is the decisive source for competitiveness.

The following in-depth case study of the German advertising industry tries to take this subject-oriented approach to understanding the interaction of economic organization, social relations and spatial structures seriously. It is structured along the three next chapters, starting precisely at the opposite side of the subject-orientation, that is, from generally portraying the structure and structural change of a globalized industry in a specific national spatial context. Particularly the structural change is then focused on as an innovation process occurring in and restructuring the national agency landscape. Finally, the emphasis is put on the individual advertising work force, stressing both its key role within the process of innovation and the ongoing dynamics in the labour market. The empirical focus is Hamburg as the present centre of the German advertising industry, and it is put into a wider context in a double way, on the one hand being seen as one element of the German system of metropolitan regions. On the other hand, it is dealt with as a rich example of a specifically urban environment for both the organization and 'functioning' of advertising and its interaction with the networks of social relations the creative professionals of the industry are embedded in.

Chapter 3

The General Perspective: Changing Advertising and the West German Space-Economy

Advertising and the Lack of an Analytical Economic Approach

Notwithstanding the fact that the rise of advertising to one of the 'privileged forms of discourse' about a lot of serious concerns of human life (Leiss et al., 1990, p.1) constituted one of the most striking occurrences of the 20th century, the advertising industry, as the business behind this phenomenon, has so far been largely ignored above all by the economically oriented part of the social sciences. That is not to say, advertising has not been subject of academic literature, at least in business economics. There is a diverse body of work very closely linked to advertising that has grown according to the degree that marketing as sub-discipline of business has increasingly obtained a scientific character. Thus, from the pragmatic standpoint of business research advertising has been acknowledged as one important element within the marketing mix, therefore being studied in relation to other marketing tools, regarding optimal use, effects on market success etc (e.g. Wells et al., 1989; Geffken, 1999).

However, more comprehensive accounts of the business, focusing on its structure and function, are scarce and generally remain limited to the macro-economic question whether or not advertising is beneficial for wealth creation. This kind of debate on the one hand resembles the way the overall service industries have been dealt with in economics, that is, starting from the traditional division of the economy in productive and unproductive parts. In this context advertising (as all services) has been considered to have only a passive or even parasitic role in the economic system of value production.[1] On the other, it is more directly derived from the peculiarities of advertising itself, given that it appears to undermine basic assumptions of economic theory, that is, first, the rational behaviour on the part of all economic actors (Dichtl and Kaiser, 1981) and, second, the sovereignty of the consumer which attributes to him the role of 'the most important decision maker in the economy' (Leiss et al., 1990, p.19). This latter aspect implies also a moral dimension for the way the sector is seen. The fact

---

1 As a nice summary of the debate on services see Marshall and Wood (1995, p.213ff.).

that the function of advertising is to 'create desires' (ibid.), that is to shift the role of the decision maker away from the consumer to the producer thereby tending to make the former more or less passive beings in the marketplace, made many important economists consider advertising as a 'morally low' (Jackson and Taylor, 1996) economic activity, 'a sort of fraud or swindle' (Dichtl and Kaiser, 1981), which exploits consumers' 'stupidity and credulity' (ibid.).

This 'moralist' attitude regarding advertising can be found in numerous variations throughout the different disciplines of the social sciences at least until the 1970s, mostly starting from radical approaches to social philosophy and adopted by cultural and media studies, parts of behaviourist psychology etc. Strongly drawing on the work of Benjamin as well as Horkheimer and Adorno and being influenced by Vance Packard's important book 'The Hidden Persuaders' (Packard, 1957) advertising was considered to be paradigmatic for the subordination of culture to the logic of capitalism thereby so to say completely 'trimming' man to the needs of the capitalist economy. Packard in addition provided the framework of depth psychology to this thinking claiming that advertising impinged upon one's unconscious both deciphering its structures and mechanisms and exploiting the knowledge about these to manipulate man's desires.

From the 1970s onwards the debates changed, due to two main reasons: On the one hand, Packard's psychological assumptions did not prove realistic. The 'big manipulator' made way for a paradigm of communication-oriented reasoning about advertising, conceiving of it as a process of 'encoding-decoding' (Hall, 1981) in which existing needs and desires are reinforced instead of being created. On the other hand, the de-industrialization of Western societies has made advertising – both directly as one part of the growth sector of business services and indirectly as an important financial vehicle for the overall creative economy – a field of growing interest for economic policy, in terms of contributing to solving economic and above all employment problems of post-industrial societies, as well as their cities and regions (Scott, 1996; Hudson, 1995; Kunzmann, 1995; AG Kulturwirtschaft, 1995, among others). In the course of this the wider theoretical focus of social science has changed, replacing or at least supplementing the culturally pessimist approach of the Frankfurt School by a variety of concepts which have provided diverse means to understand the relation between economy and culture, generally considering it more equilibrated than Horkheimer and Adorno (Lash and Urry, 1994).

In this context, some work has emerged meanwhile, focusing on recent changes in the global advertising industry by examining its globalization process (Perry, 1990; Daniels, 1995; Leslie, 1995), attempting to apply the flexible specialization theory to it (Leslie, 1997) or even exploring the reorganization of the industry as a new paradigm of sustainable (in the sense of adaptable) firm or production organization in a global and knowledge-intensive economy (Grabher, 2001, 2002).

The following chapters 3 to 5 have to be seen in line with these primarily analytical approaches to advertising as a global industry, focusing on its restructuring in the framework of the German space-economy. In this chapter we shall start with a brief conceptualization of advertising as an economic activity, from three perspectives, that is, from a macro-view of its economic function, from a micro-view of its functioning and from a more empirical view of the sector structure as well as the changes it has undergone in the course of the last 25 years which seriously questioned the unambiguous American dominance prevailing in the sector since the 1920s. Subsequently, structure and change of the German advertising industry are outlined both regarding its role in the global business and the industry's territorial pattern within the context of the national space economy. Finally the highlighted changes in this territorial pattern are interpreted as being based on an innovation process in the course of which Hamburg could emerge as a new creative capital of Germany.

Advertising as an Economic Activity – A Brief Overview

The Function of Advertising: Beyond Linear Market Communication

The most frequent way to deal with advertising as an economic activity in the literature is to classify it as a knowledge-intensive business service. This subsumption under a wider sector is certainly correct and does correspond with the official sector classification. Nevertheless it tends to unravel the complex pattern of division of labour, which is manifested in advertising and which integrates it into the wider economic system, only superficially. To understand both the role of advertising in this system and its tremendous growth in importance within this system it is thus necessary to analytically grasp advertising along with its linkages with the wider economy. In the following paragraphs we shall attempt to develop a complex understanding of the economic function of advertising by gradually widening and complicating our analytical focus.

If we start from a simple market exchange model of firms and households, advertising, unlike most of the other business services, cannot be seen as an input to production but as a mediator at the production-consumption interface (Leslie, 1995). From both a narrow economic and a radical standpoint advertising is thus a form of 'commercial capital' that as an intermediate 'organizes and adjudicates all matter of international economic exchange – for a fee' (Thrift, 1987, p.203) thereby facilitating the circulation process of capital, or 'oiling the wheels of capitalist production' (ibid., p.205). Continuing to argue in a Marxist perspective advertising growth would thus be the result of capitalists' strategies to accelerate the turnover time of capital.

Webster and Robins (1989), in their attempt to embed current changes towards the information society in a wider historical perspective, take up this line of reasoning, considering the growth of advertising as an extension of the basic features of Taylorism, 'plan and control', from the organization of industrial production to the sphere of consumption. The 'application of Scientific

Management to production' (ibid., p.331) is replaced or complemented by a 'Scientific Management of marketing' thereby permitting an intensification, extension and automation of 'the very principles of Taylorism [...] through the applications of new communications and information technologies' (ibid., p.338). Their concept of the information society thus suggests an extension of capitalists' power from the capital-labour interface to the interface between production and consumption.

A similar, but more analytical contention is launched by Sayer and Walker (1992) in the course of their differentiated conceptualization of the service economy. They classify advertising as one type of 'circulation of information' which 'accompanies the flow of commodities, money and capital [...] to weave together the multitude of production units and consumers throughout the wide social division of labour' (ibid., p.82). Following their argument the growth of advertising is thus driven by a deepening of the overall social division of labour which shifts 'the locus of social labour [...] from production to circulation, and from direct to indirect labour' (ibid., p.105). Thus advertising can be considered to indirectly help firms to achieve scale economies by saving marketing costs through the use of mass media instead of personal buyer-client transactions (Wells et al., 1989, p.11), at the same time accentuating the spatial division of labour given that replacing salespeople by mass media also enables firms to serve wider markets (ibid.). Transferred to a macro-economic argumentation this would mean that advertising reinforces the temporal and spatial de-coupling of production and consumption; put in terms of Giddens' modernization theory discussed in the preceding chapter it is one of the expert systems which drive the disembedding of social relations out of their localized contexts. Yet, in the debates on the role of advertising in the macro-economic system sketched out in the introduction above this view of an active contribution of advertising to socio-economic modernization is yet not shared unequivocally. As Leiss et al. summarize, the 'growth of advertising is correlated to the growth of the industrial economy, but whether as cause or effect is difficult to determine' (Leiss et al., 1990, p.16).

The key elements within the advertising system, which are above all responsible for its assumed dynamic of disembedding, are the mass media thereby, in the way they interact with advertising, constitute a *first complication* of the simple market model. A focus on the exchange between producers and consumers alone overlooks that the media constitute own 'players in the advertising world' (Wells et al., 1989, p.13) which do not only function as neutral message transmitters but generate a substantial share of their income from this transmission service. According to Leiss et al. advertising operates

> ... on a triple front [...] between industry and media helping to create new forms for messages about products; between industry and consumers, helping to develop comprehensive marketing campaigns; and between media and consumers, helping to do the research on audiences that led to what we know as 'market segmentation' (Leiss et al., 1990, p.12f.).

In this complex set of linkages the media do not only have their most important financial base in advertising, they also set important imperatives for the very functioning of the advertising business. Jhally in this context distinguishes 'the production of *messages* from the production of *audiences*' (1987, p.74, *original emphasises*). What is exchanged are not messages but audiences which are co-produced by the media and their viewers and readers, respectively, the price thus being a function of the quantity of viewers/readers, that is the media reach. Jhally argues that the exchange-value generated through this logic of media dominates the actual 'use-value' of the message which is to 'attach meaning' to a product (ibid., p.204).

This close economic coupling of advertising with media was for a long time also mirrored in the way the price of advertising was calculated. Traditionally it amounted to 15 per cent of the media costs. In the meantime it is dominated by a fee system in which the actual work of the agency is paid.[2] Also the subordination of advertising to the logic of the media-led exchange-value is no longer valid in the extreme sense pointed out by Jhally since media brokerage has generally split itself off as an independent institutional part. The media, however, still receive by far the lion's share of advertising investment;[3] equally, the logic of media, in the form of reach rates, programming time etc., still strongly shapes the production of advertising messages by constituting, so to speak, its 'materiality' (Paczesny, 1988), that is, the material conditions in which the messages are embedded and to which they have to adapt.

A *second complication* of the linear 'division of labour'-model of advertising lies in the simple fact that the producer-consumer interface is not only determined by the interaction of supply and demand but is decisively shaped by the competition of several producers. Advertising in this sense is a way for producers to differentiate their products and services on the marketplace in order to achieve competitive advantage. Also this differentiation can be read in two opposite ways: as 'neutral' information which helps the market to fully exploit its price-sensitiveness, or as a way to shift the differentiation criteria away from price (Wells et al. 1989, p.10). The true way in which advertising influences competition in the economy may certainly lie between these two extremes and may differ from firm to firm as well as from product to product.

Yet it appears obvious that the competition-induced role of advertising has been the main motor of the sector's tremendous growth within the last twenty five years. The general shift from seller's to buyer's markets in western economies has reinforced and accentuated competition on consumer markets thereby both

---

2   Still in 1980, 56 per cent of the revenue of German advertising agencies came from the commission system (Zuberbier, 1981). In the meantime this figure has decreased to 25 per cent (ZAW, 2001).

3   In 2000, the share of media revenue in total advertising expenditure in Germany accounted for about 70 per cent (ZAW, 2001, own calculation).

increasing the quantity of comparable products and shifting the factors of competitiveness away from purely objective criteria. The abundance of products as compared to buyers has on the one hand strengthened the position of consumers; on the other hand it has made identification as well as differentiation of products substantially more difficult. Suppliers' key tool to achieve both identification and differentiation is the *branding*, that is, the marking of the product and the attachment of meaning to the marked product (Geffken, 1999 p.136ff.).

Advertising provides a fundamental support to the branding of products and services by producing a specific, increasingly scarce good: *attention* (Schmidt, 1995; Schmidt and Spieß, 1994). The scarcity of attention is due to two crucial phenomena in contemporary western economies, that is the abundance of similar products and the abundance of communication channels, respectively the massive quantity and diversity of the media (ibid.). In this sense, advertising is a sort of 'medialized courtship dance which is to attract attention of a vis-à-vis and to direct it to a particular product' (Paczesny, 1988, p.475).

It is the nature of this vis-à-vis that implies the *third complication* of the simple advertising model outlined above. The linear perspective of an increasing division of labour and a facilitated capital and information flow presupposes a conceptualization of the consumer either as a 'passive being', manipulated through the seducing messages of advertising (Leslie, 1999), or as a rational economic actor who uses the messages as a means to improve his information base in the market exchange process. Both perspectives do not appear to meet the reality. Although the advertising industry is constantly screening the consumer, splitting him into target groups etc. via a diverse set of market research efforts, he still remains more or less a 'black box', as regards both how to reach him at all and how to persuade him to consume the supplied products. Nobody can seriously say how advertising impacts upon its audience (Schmidt and Spieß, 1994). This is, on the one hand, due to the present abundance of products and communication channels mentioned above. On the other hand, it has to do with the structure and nature of the consumption sphere itself. It is obvious that the same general process of segmentation and diversification which changed the structure of products and the media also affected people's consumption patterns. That is to say, mass markets have split up into fragmented target groups to which messages have to be directed very consciously (Scheffler, 1999). At the same time consumer behaviour itself has increasingly proven to be uncertain, obeying rather to subjective criteria than to mere product information or even manipulation (Leslie, 1999).

Thus, the role of advertising at the production-consumption interface can be considered neither as a one-sided sales catalyst nor as a facilitator of market exchange but as a means to deal with consumer diversity and consumer subjectivity by both unravelling and becoming a part of the construction of subjective personal identities.

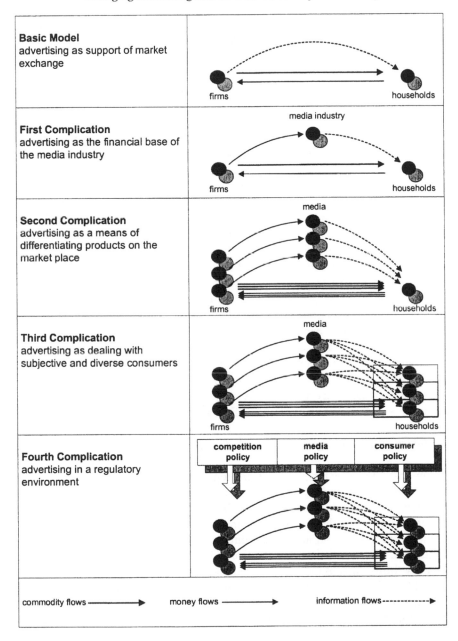

Figure 3.1   The Economic Function of Advertising – a Model of Growing Complexity

Source: Own ideas; illustration by Sybille Merbitz

The *fourth complication* of conceptualizing advertising as an economic activity is based upon the fact that it is strongly influenced by the institutional and regulatory environment affecting, on the one hand, the production of messages and, on the other, the structure of the media. As regards production, the two dimensions just dealt with, that is, advertising in a competition economy and advertising in a consumer society, are the most important starting points for regulation (Wischermann, 1995a, 1995b). Consequently the main policy objectives for advertising are prevention of unfair competition and consumer protection, the latter above all in the form of protection of consumers from particular products, such as alcohol and tobacco, or as the protection of particular consumer groups, above all children. As far as the media are concerned public regulation affects mainly the broadcasting services, via allocation of transmission licences and via provision of own services with substantial advertising restrictions.[4]

Figure 3.1 provides an overview of the outlined attempts to put advertising into a wider environment. In sum, its function in the overall economy cannot be grasped by narrowly focusing on the mediation between producer and consumer. It is true that advertising messages are transmitted at this interface. However, to understand the particular economic logic of market communication through mass media it has to be seen from a wider perspective, including the media system as the material framework of advertising messages, the consequences of growing competition on consumer markets, the increasing complexity of consumer behaviour and the institutional and regulatory environment the advertising industry is embedded in. Yet, in addition to the functional approach taken on here, one has to deal with advertising also from the micro-perspective of the main actors in order to understand its agents as well as the way they produce messages to attract consumers' attention. We shall provide a first overview in the following section.

The Making of Advertising: The Agency in the Middle of Frictions

The key institutional actor in the advertising world is the free-standing, 'classic' 'full-service' advertising agency. There are cases of in-house production by advertisers, especially in the retail sector, but usually customers hire an agency to produce and manage their market communication. 'Classic' as attribute in this context means a focus on advertising through the classic media TV, print, radio and outdoor, the so-called 'above-the-line' advertising. 'Full-service' denotes that an agency fulfils all activities of the advertising value chain from the planning and preparation of an account to media planning and buying (Ziegler, 1994). Nowadays the situation has changed in terms of both agency features: On the one hand, agencies are increasingly compelled at least to consider 'below-the-line'

---

4   How media regulation can affect advertising became visible in the process of deregulation of television in the 1980s in Europe. It did not only trigger off a tremendous growth in advertising expenditure in the eighties but also changed the media distribution of this expenditure in that TV's share rose from 16.3 per cent in 1980 to 31.2 per cent in 1996 (Howard, 1998).

advertising in their communication strategies. This is owing to the fact that the communication through the fragmented and overloaded mass media system guarantees decreasingly to reach the consumer. On the other hand media brokerage has turned into a business generally run in independent firms. As it is almost the only part of the advertising process in which economies of scale can be achieved, and as efficient media planning presupposes large investments in computer based research infrastructure it has proven to be more efficient to specialize in this part of the value chain.[5]

Thus, what contemporary full-service agencies do can roughly be characterized as 'selling ideas' to their clients (Wells et al., 1989, p.93). This general characterization of advertising activity in turn can be broken down into two core tasks: the creative development of ideas, and the account management which is so to speak to sell them, thereby serving 'as a liaison between the client and the agency' (ibid.). Figure 3.2 depicts a schematic model of the advertising value chain, emphasizing the two core tasks of the agency.

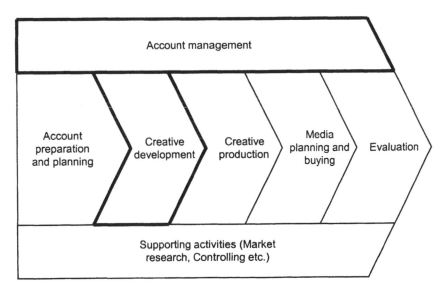

Figure 3.2   The Advertising Value Chain and the Core Tasks of an Agency

Source: Adapted from Ziegler, 1994

---

5   Yet in the case of big, internationally organized advertising networks it is normal that each network has at least one own media buying agency. This does not automatically imply that these firms collaborate always with the network's full-service agencies. Media buying contracts normally have longer lives than advertising contracts (personal interview, 2000).

Creativity and Space

The value chain model reveals that one of the agency's main jobs is to accompany the whole process of advertising campaign development in close interaction with the advertiser. In fact, the agency-client interface is not comparable with selling a product to someone. The overall strategy as well as the creative products and ex-post assessment of the strategy have to be harmonized with the advertiser's ideas, therefore requiring a high degree of feedback in the course of the development process. Another specific feature of the agency-client interface is the so-called 'conflict clause' which is an inherent part of all advertising contracts and interdicts the agency to work for any competitor of the client. This 'exclusivity' clause also expresses the closeness of the relation between advertising service firm and its advertiser; at the same time it strongly impinges on the strategic orientation of an agency given that it is hindered in specializing on one particular product group.[6]

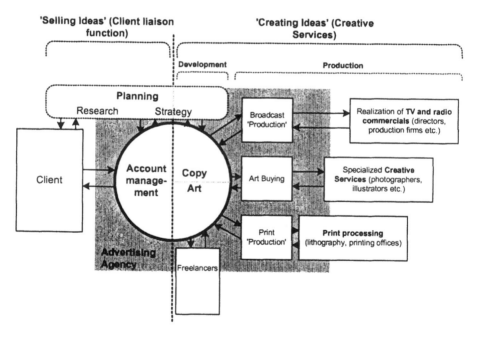

Figure 3.3   Creating and Selling Ideas in and around an Advertising Agency

Source: Own illustration

This intensity of collaboration and, therefore, of co-ordination within the development of advertising campaigns does not only refer to the relation between agency and advertiser but is also reflected in the organisation of the production inside the agency and at the various interfaces between the agency and its suppliers of diverse production inputs

---

6   See Wells et al, 1989, p.104 as well as Perry, 1990, p.39 for an encompassing explanation.

which is depicted in Figure 3.3. The figure is not an illustration of the organization of an advertising agency but aims to summarize how the functional nexuses within and around it work in the development and production of campaigns. What appears crucial is that the double-faced core of creative development, with art and copy as key occupational profiles, and account management functions as a 'hinge' through which the diverse productive inputs are harmonized with the communication strategy developed in close co-operation with the client.

Yet the fulfilment of the agency core's 'hinge function' in the process of campaign development is by no means a pure co-ordination task. The double role of the agency to both create ideas and sell them to business clients is far from being without frictions or ambivalences. Thus, besides co-ordinating the different inputs to a campaign and binding them together to a consistent strategy, the core of the agency has to bridge a fundamental cultural barrier between the creatives' 'artistic thinking', on the one hand, and the demands coming from the client side and the market research, on the other. How this harmonization between different 'action frameworks' (Storper, 1997) works, depends on several variables: the client's general 'taste' in terms of the advertising style, the situation on product markets, the advertising strategy of competitors etc. Additionally, the relation between the business part and the creative part of an agency has changed in the course of the last fifty years. Whereas in the period of growing mass markets after World War II the creative development and production – as one department of a vertically integrated, taylorist agency – was clearly subordinated to the marketing research-led account management, the diversification of demand and the growing uncertainty of consumer behaviour from the 1970s onwards has strengthened the role of creativity and made the barrier within the agency more permeable.[7]

According to Lash and Urry (1994) this resulted in a product innovation of a more creative, 'entertaining' advertising style which was in turn encouraged and supported by a crucial process innovation managing 'to reconcile the creative and marketing research approaches to advertising' by putting them together and co-ordinating them through the function of 'account planning' (ibid.). 'Planning' as an integrated strategy development has in the meantime generally been included in the advertising process, however in diverse ways and following diverse approaches and philosophies, normally not institutionally integrated as a 'third column' of the agency structure (Lange, 1999, p.20). It can be accomplished through the interaction of account management and creative development itself, in this context either in a top-down approach in which the creative work follows the general strategy or vice-versa, that is, as a strategy built around an idea from the creative work (Schmidt and Spieß, 1994); it can function as a service department which can be included in this interaction either voluntarily or compulsorily, serving to refine the strategy in terms of 'effectiveness' or 'efficiency' etc. In other words, although

---

7   We shall discuss the background of this change in the following section on the general structural transformation of advertising as well as throughout the following two chapters, given that we consider the restructuring of the German advertising industry to be based upon this 'creative turn'.

the importance of strategic account planning is generally acknowledged its actual implementation is still very diffuse and differs from agency to agency.

In contrast to that, agencies are very similar in the way they organize the creative production for the different communication channels. Usually the actual production is not carried out inside the agency but given to specialized firms. Unlike suggested by transaction cost economics that considers more complex activities to be generally carried out in-house, the likelihood of a production activity in advertising being conducted by a free-standing specialized supplier tends to increase with the complexity of the input. Thus, print processing or at least the preliminary stages of print production (lithography etc.) may be part of the agency's activities if economies of scale can be achieved in this area, due to a specialization in certain print media such as catalogues, brochures etc. or due to an agency size that implies a critical mass of print products. Moreover, the increasing digitalization of the creative work has strongly affected the pattern of division of labour in print advertising since the final art work and layout is generally done on the art director's screen, print production thus merely being a technical realization of an already finished ad.

At the other extreme, TV- and radio production are nearly always conducted by independent enterprises. This general externalization is mainly due to four reasons: First, audio-visual production requires a very specific technological know-how different from the traditionally print-oriented advertising. Second, it also presupposes large investments in specific studio and post-production equipment. Third, each commercial is normally produced by a myriad of independent and self-employed actors, from the director to the camera assistant, all co-ordinated by the production enterprise. To carry out this co-ordination task would certainly overstrain the agency. Fourth, and most importantly, the interaction between agency, production firm, director, actors and other participants in the production process is not a simple supply of an ordered product. On the contrary, although the script and storyboard[8] are generally written by the agency, it is important for the quality of the film or radio-spot to exploit the 'genuine creative input' (Grabher, 2001) of the suppliers as much as possible (Schmidt and Spieß, 1994).[9] It is just at the frictions between different creative agents that the best results are produced.

Inside the advertising agency the actual production of advertising is reflected in three different functions which rather constitute interfaces between creative development and external production than being own producers. In this sense, they rather are 'traffic' departments than production units in that they manage the process of co-operation between the agency's creative core and the suppliers of creative services, in terms of organization, quality control, deadlines, copyright

---

8   The storyboard is a visualization of the script which is made for presenting a
     commercial to the client.
9   Grabher (2001, p.368) compares the co-operation of advertising agencies with
     advertising film producers to improvisation in jazz. A general theme is interpreted by
     the single musicians, results are unpredictable, 'errors' are considered as 'a source of
     learning', 'leadership is rotated' etc.

regulations etc. In bigger agencies, these production/traffic functions are accomplished by three centralized departments in which the specific knowledge about each input channel is concentrated and scale economies can be achieved.

A very important linkage to the exterior of the agency is not managed through specialized departments but by the creative core personnel itself: the interface to a pool of freelancing art directors and copywriters whose labour force is temporarily used to flexibly react to changing volumes of work to which the level of internal labour force cannot adapt. These freelancers frequently even take on the role of quasi-employees who work regularly in the creative development of one agency without officially being employed by it. Their quantitative importance for the overall sector is enormous: According to the German Microcensus in 1997 more than 27 per cent of people working in the advertising industry were classified as 'self-employed persons without employees'.[10]

The illustration leaves out the two parts of the market communication process outlined at the outset of this section, that is, on the one hand, media planning and buying where the agency's role can differ from client to client and from account to account and, on the other, 'below-the-line' market communication such as direct mailing, sponsoring, sales promotion, public relations, event marketing etc. In this diverse and strongly growing field the co-ordination function can vary between the lead (full-service) agency and the client himself. Particularly big and internationally oriented agencies however tend to position themselves as 'total' or 'integrated' communication groups which are able to master the whole range of advertising channels, therefore at present increasingly aiming to include non-classic advertising in their activities.

Thus, advertising appears to have become an even more complex co-ordination task than outlined in the figure. Future full-service agencies' success may substantially depend on their capacity to orchestrate clients' demands, different communication channels and the variety of inputs by specialized producers. Nevertheless, co-ordination will never be a pure 'streamlining' of different elements in the process of strategy development for market communication. On the contrary, what distinguishes advertising from other economic activities is that it mediates between different 'worlds of action', that is between the artistic world of ideas and the business world to which these ideas have to be sold, in this context handling the existing frictions in a complex way rather than simply abolishing them. We shall in the course of the three empirical chapters argue that the challenge to bridge the different worlds without abolishing the frictions between them is one of the crucial features of the contemporary global advertising industry and that this challenge substantially influences the structure and organization of the whole sector.

---

10 Own calculations based on the Microcensus scientific use file, 1997.

The Structure of Advertising: The Rise of the 'Second Wave'

The description of the function and the making of advertising in the course of the two preceding sections have shown that advertising, both on the macro-level of the economy and on the micro-level of the advertising process, is part of a multilateral set of relations and institutions which substantially shape its performance. Discussing the structure of the industry and above all the structural change it has undergone within the last about 25 years consequently requires an equally complex approach, taking at the same time changes in the agency structure, the media structure, and the market structure (both in terms of advertisers and the consumer market) into account. To begin with, we shall yet focus on the macro-geographical structure of the global industry since it provides in our judgement the best impression of the industry's structural character and of the restructuring process it has gone through in the course of the so-named 'second wave' of global advertising.

The geography of the global advertising industry can be read from two different perspectives: from a market standpoint, that is, in terms of a geography of advertising expenditure and from a supply standpoint, that is, in terms of the geography of 'production'. As regards expenditure, advertising more or less reflects the inequality pattern of the world economy, with North America and Europe accounting for about 70 per cent of the world's total spending in the sector in 1996, together with Asia/Pacific even for 93 per cent (NTC Publications, 1998). This pattern is repeated in a more differentiated way when breaking it down to the national level, with the United States (36.2 per cent), Japan (15.7 per cent) and Germany (8.8 per cent) at the top of the ranking (ibid.). Yet the polarization used to be stronger. In 1970, the share of the US market in advertising expenditure amounted to 62 per cent of the world (Perry, 1990); still in 1987 almost half of the global spending (47.7 per cent) was done in the United States (NTC Publications 1998). Thus, from a market standpoint, it seems legitimate to argue that the history of modern advertising is to a very large extent linked to the success of American 20th century capitalism (Grabher, 2001; Mattelart, 1991). The last quarter of the century has seen a sort of up-catching modernization on the part of the rest of the world, with comparably moderate growth rates in the advanced economies and substantial growth rates in emerging economies (Daniels, 1995). However, advertising expenditure has largely remained a function of the economic strength of national economies.

As regards the supply structure the geographical pattern, despite revealing the same American dominance in an even more radical manner, is more complex in that it has not been economic power alone that shaped the structure of global advertising production. In other words, American advertising networks have been dominating the global business throughout the 20th century and continue to be the key players in it. However, two main additional aspects have to be considered in this context: First, the positive relation between the size of the national advertising market and the international importance of national advertising production valid for the US can only in a limited way be translated to other nations, if at all. The most obvious example for a disarticulation of the importance of the advertising market and the national advertising production system is Germany, the particular logic of which we shall try to unravel in the following empirical part of this chapter. Second, the total American predominance in global advertising supply has been seriously challenged in the course of the last twenty years. In the following paragraphs we shall discuss the mechanisms behind this fundamental change from an 'imperial model' (Mattelart 1991) to a more global and multipolar configuration.

Historically the internationalization of American advertising agencies was closely linked to the internationalization of big American industrial firms. Agencies were simply following the requirements on the part of their main clients. In Grabher's words:

> The advertising industry [...] followed client industries not just in organizational terms but also literally, that is, in terms of the geography of production. In the wake of the large Fordist companies, US advertising agencies moved abroad, again in response to the needs of internationalizing clients to deal with a single agency world-wide (Grabher, 2001, p.351).

Thus, in accordance with the transnational activity of big US consumer brands, American agencies began to cross their national borders very early. J. Walter Thompson as pioneer in this context even established its first office abroad in London in 1899, accompanying its main client General Motors to European markets. From the late 1920s onwards and in a second movement after World War II similar pairs of agencies and clients followed this example (Mattelart, 1991). This international network-building on the part of American agencies until the 1980s actually constituted global advertising: In 1981, for instance, American firms accounted for 95 per cent of the total turnover made by all advertising agencies world-wide outside the borders of their home countries (Ziegler, 1994, p.81).

At the latest from the early 1980s onwards this unilateral pattern, widely discussed as the 'first wave' of international advertising (Lash and Urry, 1994; Leslie, 1997; Grabher, 2001) has changed dramatically. Until 1991 the US share in world-wide turnover abroad decreased to about 51 per cent (Ziegler, 1994, p.81). Mainly the UK succeeded in assuming a significant role in the world of advertising, shifting its share from 0.5 per cent in 1981 to nearly one third ten years later (ibid.). At the same time the degree of internationalization in the sector has grown dramatically during the decade: world-wide foreign turnover increased by 480 per cent between 1981 and 1991.

However it is not only this operational dimension of cross-border advertising activity in which the industry has become more global since the 1980s. Also, and more importantly, the 1980s also opened the advertising industry to the logic of financial markets, thereby substantially reorganizing the traditional pattern of advertising firms through a series of mega-mergers, respectively through the establishment of big mega-holdings which united several internationally organized advertising networks, bringing about a massive concentration of the overall business.[11] With this holding-structure which had been pioneered by the Interpublic Group in the early 1970s (Mattelart, 1991) the agency networks aimed at overcoming the growth-limiting effects of the conflict clause (Ziegler, 1994). The different networks work independent from each other, even competing among one another in terms of the operational business, but they are submitted to the same financial logic of one public company. In this sense the tremendous increase of the internationalization of UK advertising owes to a large extent to this financial dimension of globalization: The lion's share of the UK's growth was based upon hostile take-overs of big US agencies by UK-based holdings, with WPP and Saatchi & Saatchi as prime examples.

Yet it would be seriously misleading to describe the recent change in the global advertising industry purely in the metaphors of growth and concentration, respectively as a subordination of advertising to the global financial market. The relative loss[12] of importance on the part of US-networks also reflected a deeper crisis of the 'first wave' of international advertising which was again linked to the general transformation Western societies have undergone since the 1970s. This 'crisis' resulted, on the one hand, in an increasing instability of client-agency relationships formerly characterized by mutual long-term commitments and common success stories (Leslie, 1997).

On the other hand, it revealed the weakness of the traditional, big, factory-like American advertising firm that proved to be little apt to the changes in the consumer market. European, and particularly British agencies seemed able to provide the more entertaining and creative advertising style required by 'the growing opposition, scepticism and resistance on the part of the consumer towards advertising' (Leslie, 1997, p.1022). In contrast to that American agencies' campaigns appeared too 'bombastic' (Lash and Urry, 1994) and 'conservative' (Leslie, 1997) to reach the increasingly sceptical audience. Shapiro et al. nicely describe how British creatives saw their task in taking consumers' scepticism seriously, thereby considering their relation with the audience as being based on a bargain, paying, so to speak, for people's attention by entertaining them:

> ... from the late 1960s a characteristically 'British' sort of advert was created that was not bombastic, declaratory or literal as the American adverts typically were, but was

---

11 In 2000, the three biggest communication holdings Omnicom, WPP and Interpublic covered about 39 per cent of the world-wide gross income (Advertising Age, 2001). With the acquisition of True North, the ninth largest group, by Interpublic in the beginning of 2001, the share of the top three has risen to more than 40 per cent (*Horizont,* 12/2001). See, as a documentation of the concentration process, Mattelart, 1991; Lash and Urry, 1994; Ziegler, 1994; Leslie, 1995; Grabher, 2001, among many others.

12 Note that also US-agencies were able to increase their abroad turnover by 260 per cent (Ziegler, 1994). In addition to this, American ad organizations still are the most important players in the global business with five agencies out of the top ten in the world (Advertising Age, 2001).

based on a more humorous, self-deprecating, ironic and rather more subtle style. This has been seen as bargain struck, initially between British advertisers and audiences but since widely imitated elsewhere, to consume the ad in exchange for being amused, or dazzled, rather than hectored or patronized (Shapiro et al., 1992, p.191).

Schmidt and Spieß (1997) propose another, more supply-led explanation of advertising change towards more creativity: The increasing saturation of markets and homogenization of product quality has shifted the focus of competition from products to communication. This is in some cases even mirrored in the actual campaign development in that it is founded on the analysis of competitors' advertising, chiefly aiming at differentiating the communication strategy (von Matt, 2002).

The overall shift in favour of creativity having affected global advertising from the 1970s onwards, as a movement parallel to the massive globalization during the 1980s, gave rise to a whole body of small, but fast-growing firms which were placing their emphasis clearly on creativity thereby even challenging the traditional agencies in their classic field of globally marketed consumer brands.[13] The new period, which, while having earlier roots, began to transform the overall business seriously from the 1980s onwards, has been labelled – in analogy to the period of unrestricted American dominance – as the 'second wave' of advertising (Lash and Urry, 1994; Leslie, 1997; Grabher, 2001). It has replaced the 'first wave' by a more complex industry pattern, in which processes of global concentration interact with a decentralizing tendency encouraged by the increasing importance of creativity. The 'second wave' of international advertising brought about further changes in the wider field of advertising, that is, in the media system, in the structure of clients, and, importantly, in the locational pattern of agencies, both on a global level and on the level of different national space-economies.

Concerning media, television has become the lead medium for market communication.[14] Regarding the client structure the main carriers of the first wave, that is, the big American fast living consumer brands like Kellogg's, Kraft, Coca-Cola, Pepsi, Marlboro etc. (Mattelart, 1991), despite maintaining important positions in the global rankings of advertising expenditure, were no longer the main vehicles of growth. At first the car market, and later further consumer durables, financial services, telecommunications etc. widened the spectrum of clients and confronted agencies with new tasks. Table 3.1 depicts the top ten

---

13 The most famous case in this context was the shift of a 'significant proportion' of the Coca-Cola account 'from McCann-Erickson Worldwide [...] to a Hollywood based talent agency, Creative Artists of America (CAA)' in 1993 (Leslie, 1997).

14 On a global level the share of television in total advertising expenditure between 1987 and 1996 rose from 30.1 per cent to 38.1 per cent (NTC Publications, 1998). In Europe it nearly doubled from 1980 to 1996, that is, from 16.3 per cent to 31.2 per cent (Howard, 1998). The data for Germany are different since the media breakdown includes categories outside the classic media such as postal advertising. Here the television share grew from 8 per cent in 1984 (Schmidt and Spieß, 1994) to 20 per cent in 2000 (ZAW, 2001), yet obviously having reached a saturation point in 1997 (ibid.).

Creativity and Space

spenders of global advertising in 1989 and 2000, showing above all the tremendous growth of car advertising. The big groups Unilever and Procter & Gamble remain on their top positions particularly in terms of expenditure outside the US, but the overall pattern has become more varied.

Table 3.1 World Top Advertisers 1989 and 2000 by Global Ad Spending (Mill. US$)

Rank	1989 Advertisers	Total ad spending	Outside the U.S.	2000 Advertisers	Total ad spending	Outside the U.S.
1	Procter & Gamble	2,713	934	Procter & Gamble	4,152	2,610
2	Philip Morris	2,502	430	General Motors	3,979	1,028
3	Unilever	1,744	1,140	Unilever	3,664	2,967
4	General Motors	1,688	324	Ford	2,323	1,127
5	Nestlé	1,143	537	Toyota	2,135	1,345
6	Mc Donald's	934	160	Nestlé	1,886	1,560
7	PepsiCo	913	127	Volkswagen	1,714	1,290
8	Ford	871	269	Coca-Cola	1,579	1,176
9	Kellogg's	824	212	Peugeot-Citroën	1,004	1,004
10	Toyota	740	322	Fiat	990	988

Sources: http://www.adageglobal.com/cgi-bin/pages.pl?link=498, 7/1/2002; Baums, 1991

As regards the reshaping of the agencies' locational pattern, both centralizing and decentralizing tendencies can be identified. On the global level, in analogy to the relative loss of American agencies, New York's Madison Avenue, which had been the 'undisputed centre' of the first wave (Grabher, 2001), had to experience the emergence of a serious competitor, London, as starting point of the top agencies of the new era and 'laboratory' for a new advertising style (ibid.). On a European level the rise of London led to a significant re-concentration of the activity in the UK capital, to the detriment of other, previously dominant European cities, such as Paris, Stockholm, Amsterdam or Düsseldorf, at least in terms of top agencies (Daniels, 1995).

But there were also significant changes in the different national space-economies. In the United States it was no longer a precondition for successful advertising to locate in Madison Avenue. On the contrary, newly emerging small creative agencies explicitly avoided the former centre in order to signal 'to the advertising community' to be 'outside the mainstream' (Leslie, 1997, p.1030). California and Oregon, as well as locations inside New York, but outside Madison Avenue, became the sites of the 'second wave' community in the US.

In the meantime, the path-breaking transformations characterizing the 1980s have lost pace and even partly the strength of their impact. The key players of the first wave have been able to maintain an important position within contemporary global advertising whatever the group they belong to. Also Madison Avenue has by no means disappeared from the advertising map, but hosts two of the headquarters of the top three holdings, as well as seven of the world's top ten agency brands (Advertising Age, 2001). As we put above, the rise of television as key medium of the second wave appears to have reached a limit also.

However, the basic structural shifts in advertising production, that is, the massive concentration of agencies in shareholder-oriented holdings, on the one hand, and the need for creativity to attract consumers' attention have pervaded the overall business, and have influenced also its locational pattern. We shall now shift our focus to a very interesting national case in this context, due to both the specific structure of the space-economy and a pronounced disarticulation between the size of its advertising market and the poverty of the domestic agency structure: Germany.

German Advertising and the German Space-Economy

External Control of a Large Market: German Advertising

It appears surprising that the third biggest national advertising market in the world (ZAW, 2001) and the country with the highest per capita purchasing power in the European Union has not been able to develop an internationally successful advertising industry. In the most recent ranking of the world's top advertising organizations there are only two German agencies among the top 100, the first of them ranking at no. 51 (Advertising Age, 2001). Even the home market is basically dominated by the German subsidiaries of international network enterprises. Among the biggest twenty advertising agencies in Germany in 2000, only five had their operational headquarter within the country, one of them being completely owned by, another one having sold a 35 per cent share to an international holding. Table 3.2 illustrates this dominance, particularly highlighting the supremacy of the top three major groups.

Yet this relatively weak performance on the part of the German advertising industry both on its home market and particularly in an international context can be explained by its historical evolution which brought about a specific situation in terms of *client structure*, *media regulation* and *agency system*, thereby handicapping the advertising sector to grow in an endogenous way.

Creativity and Space

Table 3.2 German Top 20 Advertising Agencies Ranked by Gross Income (2000)

Rank 2000	Agency (-Group)	Global Holding (Rank world-wide)	Gross Income in Mill. Euro 2000
1	BBDO Group Germany	Omnicom* (2)	270.087
2	Grey Global Group Germany	Grey (8)	136.515
3	Publicis Group Germany	Publicis (6)	133.125
4	McCann-Erickson Group Germany	Interpublic* (3)	91.208
5	Ogilvy & Mather Group Germany	WPP* (1)	86.962
6	Young & Rubicam Group	WPP* (1)	77.902
7	Scholz & Friends Group	Cordiant (10)	71.649
8	Springer & Jacoby Group	35% True North (9)	63.312
9	DDB Group Germany	Omnicom* (2)	61.529
10	J. Walter Thompson	WPP* (1)	58.801
11	Michael Conrad & Leo Burnett	Bcom3 (7)	53.712
12	FCB Germany	True North (9)	53.519
13	TBWA Germany	Omnicom* (2)	51.103
14	Lowe Communication Group	Interpublic* (3)	46.091
15	ServicePlan Group	-	44.994
16	D'Arcy	Bcom3 (7)	44.225
17	Citigate Group Germany	Incepta (20)	41.249
18	Heye & Partner	Mino. Omnicom* (2)	35.194
19	Jung von Matt	-	34.755
20	Euro RSCG Group	Havas (5)	28.431

* Top 3 Holding

Sources: *Werben und Verkaufen*, 13/2001; Advertising Age, 2001, p.s18

As regards *clients*, the capital-goods orientation of the German manufacturing industry hindered the establishment of a strong home market. Particularly the fast living consumer goods, which constituted the big brands through which advertising grew after World War II, played a minor role in the national industrial structure. Thus, Germany lacked a critical mass of a client industry through which a domestic advertising market could emerge as starting point for internationalization.

The limitation of the home market also occurred through the sharp regulation of advertising in the German *media system* – particularly as regards public TV, in which the total transmission time of commercials is 20 minutes per day, the time after 8 p.m. as well as Sundays and other holidays totally excluded (Schmidt,

1999).[15] This restrictive policy was caused by a generally sceptical attitude to advertising in Germany, but also by the negative experience made with nazi propaganda which had made extensive use of the mass media (Ziegler, 1994).

The German *agency structure* was originally not dominated by a service-orientation. Advertising agencies had their origins as in-house offices within the big publishers' groups, therefore functioning chiefly as their 'fund-raising agencies'. Consultancy and creative services were generally accomplished by the clients themselves (ibid.). The orientation towards free-standing business services for clients from different sectors was so to speak 'imported' with the establishment of international corporations on the German market, being accompanied by their own advertising agencies. McCann and J. Walter Thompson, the early movers of the first wave of international advertising, together with their clients Standard Oil (Esso) and General Motors, respectively, both set up in Berlin in 1928, were the first examples for such an import of a new type of service enterprise. As the only exception from this importation pattern LINTAS, established in Hamburg in 1929, constituted a quasi-local agency since it was set up as a department of the Unilever-Group, a merger of the Dutch Margarine Unie and the British Lever Bros., the former of which had previously bought out a Hamburg based producer of vegetable oil (Läpple, 1989).[16]

Few years later, Nazi dictatorship brought about a significant throwback for this young service industry, creating, already in 1933, an advertising law through which the whole sector came under strict control of the regime (Zuberbier, 1981). Only the 1948 monetary reform after World War II made the agencies resume their activities. But also then the growth of agencies occurred very slowly: In 1970 only 50 full-service agencies were estimated for the whole country (Kellner, 1995). In addition, the pre-war import pattern of agency structure was maintained and even reinforced in the growth of the German economic miracle of the 1950s. A whole body of mainly US advertising agencies either followed the model of close 'service-client nexus' or took over local agencies thereby bringing about the general hegemony of the vertically integrated full-service agency (Schröter, 1997).

Nationally owned agencies at that time were either new foundations after World War II or the 'rest' of the old publisher-oriented media space buyers that had dominated the domestic part of the sector before 1933. The main innovations had however been introduced by the American network affiliates. With their high degree of methodological professionalism they were clearly dominating the sector (ibid.). That is to say, national advertising players on the West German post-war market from the very beginning had a very fragile position, strongly determined by their co-existence with the American standard model. They were, on the one hand, forced to adapt their development to the new, mainly organizational standards set by international players. On the other hand, they had to look for local market niches

---

15 This regulation has been established in 1962 and is still valid for the public channels.
16 Actually LINTAS was a subsidiary of the British Lever, also expressed in its name **Lever INTernational Advertising Services**.

left by them. Examples of successful fast growing firms were on the one hand rare, and, on the other, likely to be bought out by international agencies.

As regards location, Frankfurt and Düsseldorf inherited Berlin's role as German advertising centres at that time since they had been able to attract the German headquarters of the corresponding multinational firms (Zuberbier, 1981). Hamburg remained so to speak the third player due to the presence of LINTAS, for a long time the biggest among the international network agencies in Germany, as well as due to the existence of port-related consumer goods industries (e.g. petrol) which made many of the international networks open branch offices in the city. Thus, in comparison with other countries, the spatial structure of the (West-)German post-war advertising industry revealed and still reveals a quite decentralized pattern. Yet, in comparison with the structure of other service industries sectors in the German space-economy it appears rather polarized (Läpple and Thiel, 1999).

Advertising kept growing in Germany until 2000, through the global economic crisis and the crash of the 'new economy' since the middle of 2001 having now experienced a period of stagnation for the first time (ZAW, 2002). With the establishment of private TV in 1986 one important barrier to its expansion fell; along with the unification it made Germany take the UK's place as third biggest advertising market in the world in 1990 (Howard, 1998). Advertising expenditure amounted to 33.21 billion Euros in 2000, that is, 1.63 per cent of GDP (ZAW, 2001). It continuously achieved higher growth rates than the total national economy during the 1990s (ibid.). According to the German Microcensus, employment in the advertising sector amounted to 155,191 in 1997, that is, an increase by 16.3 per cent as compared to 1991.[17] The calculations by the German Advertising Observatory (ZAW, 2001) which also include advertising professionals in the media and client industries as well as the employees of supplying sectors estimate a volume of 361,000 employees in German advertising in the year 2000, as compared to 350,000 in 1996 (ZAW, 1997) and 359,000 in 2001 (ZAW, 2002).

What these figures do not highlight are the qualitative transformations behind this growth. In the 1970s the framework for advertising in Germany changed parallel to other western countries. The changes of the whole industrial world made also the German advertising market be caught by the 'second wave', implying the need for a more creative and entertaining advertising style. This general trend was reinforced by the restructuring of the German media landscape, in that the growing quantity and diversity of media supply widened the spectrum of communication channels, but at the same time complicated being perceived by the consumer audience at all (Jung, 1999).

---

17 Own calculations based on the German Microcensus scientific use file, 1991 and 1997. In the following sections we shall use different figures since our analysis will be based on West Germany, including Berlin, in order to avoid as much as possible distortions through the radical transformation in the eastern part of the country.

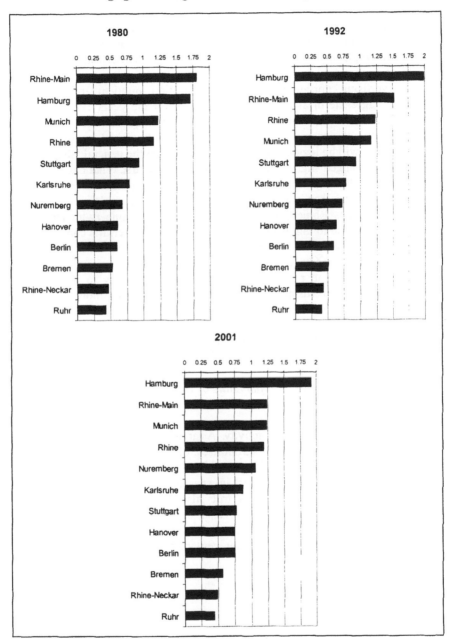

Figure 3.4 **Ranking of West German Metropolitan Regions – Location Quotients in Advertising (1980, 1992, 2001)**

Source: Employment statistics, own calculations

Of course, the 'second wave' of advertising also affected the locational pattern in Germany. Hamburg gained significant importance as new national growth centre, nearly replacing Frankfurt and Düsseldorf particularly in terms of the employment dynamic. Figure 3.4 illustrates this shift of the 'centre of gravity' in German advertising northwards, using the location quotient, based on the employment aggregate of all metropolitan regions, for the years 1980, 1992 and 2001 as indicator, that is, relating the weight of advertising with the weight of the overall economy of one metropolitan region.[18] Both 'weights' being equal the location quotient will be logically unity. One can see that the general 'shape' of the figure has only changed slightly during the last twenty years. What has changed is the order between the regions, with Hamburg clearly taking the lead between 1980 and 1992, and since then maintaining the distance more or less, whereas the previous lead regions Rhine-Main (with Frankfurt) and Rhine (with Düsseldorf and Cologne) continue to fall behind. What is surprising is Berlin's constantly low position. Set in relation to the overall regional employment the advertising cluster in the new capital still seems to lag behind.

So as not to come to precipitate conclusions we shall try to illustrate and unravel the logic of advertising in the German space-economy more profoundly, placing the emphasis on the changes the sector has undergone in the last about twenty years. In this context we shall continue to pursue and deepen the interregional analysis just started, focusing on the decentralized system of metropolitan regions which is characteristic for the German space-economy.

The German Space-Economy: The Ambiguity of Urbanness and Hamburg's Rise as Creative Capital

Advertising is an essentially urban activity. A large portion of the global business is done in a handful of large urban metropolises, of 'Global Cities' which function as administrative hubs in world-wide advertising networks. But also below the sector's geographical top level big urban spaces host most of the successful agencies. In general terms this holds true also for the territorial pattern of advertising in Germany. As Figure 3.5 reveals, in 2001 more than 46 per cent of advertising employment is concentrated in the cities with more than 500,000 inhabitants, as compared to about 23 per cent of total employment these cities account for.

---

18 The metropolitan regions are a spatial aggregate of the NUTs-3 level which is constituted by the German rural districts (Landkreise). They include both urban cores and fringes, except in the case of Berlin where until 1992 the pre-unification borders of the western part are used. For a discussion about selection and delimitation of the regions see Pohlan, 2001. The years 1992 and 2001 have been chosen for representing the maximum employment levels in West Germany within the available time series.

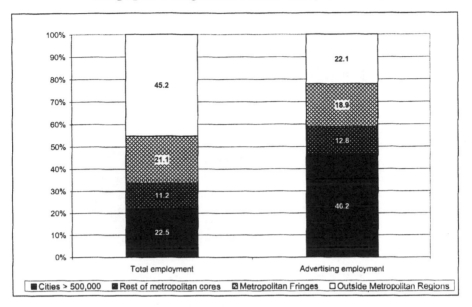

Figure 3.5   Share of Total Employment and Advertising Employment by Territorial Categories – West Germany and Berlin (2001)

Source: Employment statistics, own calculations

Looking at the changes this pattern has undergone in the last 20 years, the big cities on the one hand seem to have lost their undisputed dominance. Figure 3.6 shows their dramatic decrease in relative importance above all during the 1980s (from more than 52 per cent in 1980 to 46 per cent in 1992, from then on remaining relatively stable[19]). However, this shift has to be considered in the light of a general growth in total advertising. Referring to the total of our spatial aggregate West Germany plus Berlin advertising employment nearly quadrupled between 1980 and 2001, from about 37,000 to more than 131,000 employees. Figure 3.7 depicted the breakdown of this absolute growth along the territorial hierarchy: In absolute figures big cities still display the strongest growth: More than 40,000 of the almost 95,000 new advertising jobs have been generated there.

---

19 This stability is however also relative given that the share had decreased until 1999 to 43.7 per cent and returned to decrease with the crisis of the 'new economy' after 2001 (44.9 per cent in 2002).

Creativity and Space

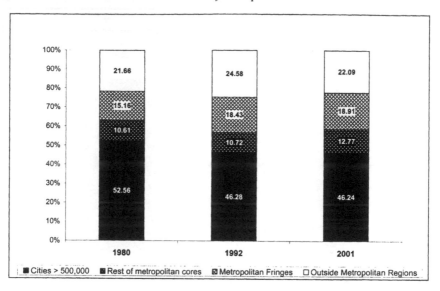

**Figure 3.6   Share of Advertising Employment by Territorial Categories –
West Germany and Berlin (1980-2001)**

Source: Employment statistics, own calculations

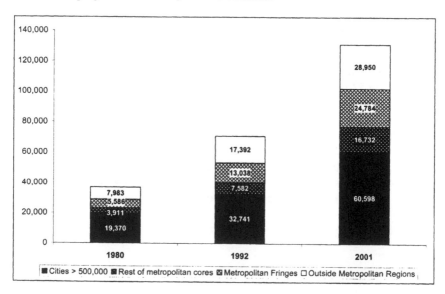

**Figure 3.7   Absolute Advertising Employment by Territorial Categories –
West Germany and Berlin (1980-2001)**

Source: Employment statistics, own calculations

Thus, as regards the role of the urban as an environment of advertising activities the development within the last about twenty years is ambiguous, revealing both a strong absolute growth and a slight relative loss on the part of the big urban centres in the German space-economy. The relative loss could, at first glance, be explained as market driven: A growing demand by the overall economy tends to make the market areas shrink in which advertising remains cost-effective (Maillat and Vasserot, 1988, p.172). This argument would be at least partly supported by the strong relative growth of metropolitan fringes – that is, following the general tendency of industrial location in the West German space-economy (e.g. Bade and Niebuhr, 1999) – and the smaller core cities within the metropolitan regions. The growth of the latter can be seen as a kind of locational trade-off between proximity to clients and the need for urbanness.

This pure market focus is however inconsistent with the changes among the metropolitan regions visible in Figure 3.4. Hamburg's advertising sector, for instance, already developed comparably well in the first half of the 1980s when the city's economy was experiencing a severe crisis, when thus the potential local market developed badly. Thus one may assume that the parallelism of two processes within advertising, that is, of market growth on the one hand and the qualitative change towards a more creative style on the other, has diverse impacts on the industry and its territorial pattern, given that the advertising sector and its market is in itself heterogeneous and at least partly segmented. Roughly put, it can be split into agencies working for regional small and medium sized enterprises, and agencies providing advertising for the big brands on the national market, on the other. In the case of the former, the argument of a growing demand and consequently shrinking market areas may certainly apply, and the suggestion of a centrifugal dynamic supported by an equally centrifugal tendency of manufacturing and service SMEs is probably right. In the case of the latter, however, the dynamic is just opposite. The number of brands advertised nationally through mass-media has increased dramatically during the last twenty years[20] causing an evenly tremendous growth of advertising in agencies serving the national market. A regional agency-client nexus might be possible and sometimes even helpful, but it cannot be considered as an essential condition. The continuing polarization between the metropolitan regions even suggests that, on the level of national and international markets, these intra-regional linkages are only secondary criteria in terms of the agencies' location.

This polarization is also visible when focusing on the big cities with more than 500,000 inhabitants alone. As Figure 3.8 reveals nearly 80 per cent of their advertising employment is concentrated in only five of the twelve urban centres. Comparable to the findings on the regional level in Figure 3.4 Frankfurt and Düsseldorf have substantially lost importance, whereas Hamburg, Munich and Berlin have increased their share, the latter particularly thriving during the 1990s and meanwhile representing the second biggest advertising cluster in the country

---

20 According to Nielsen/S+P the number of brands advertised on the German market increased from 32,850 to 53,272 between 1980 and 1998 (Jung, 1999, p.48).

although still lagging behind when related to its actual size (Figure 3.4). Thus in an interregional perspective, given that Berlin, Hamburg and Munich constitute the major urban centres of Germany, the advertising industry has obviously moved towards more urbanness.

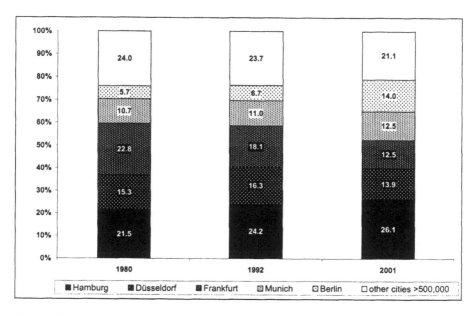

Figure 3.8   The Distribution of Advertising Employment in West German Urban Centres (1980, 1992, 2001)

Source: Employment statistics, own calculations

The different trajectories of the five top advertising centres illustrated in Figures 3.9 and 3.10 confirm this move, yet revealing that, while advertising employment Hamburg has constantly developed above the average since 1980 (except during the 'new economy' collapse after 2001), the extraordinary growth in Munich and (particularly) Berlin is almost exclusively a matter of the 'new economy' boom of the end of the century. Whether this is only a temporary phenomenon of the 'bubble economy' or a more persistent trend cannot be assessed yet. More recent accounts on employment changes in the IT and media industries as a whole at least show that also during the crisis since 2001 the three biggest centres have performed better than the rest of West German urban cores (Frank et al., 2004).

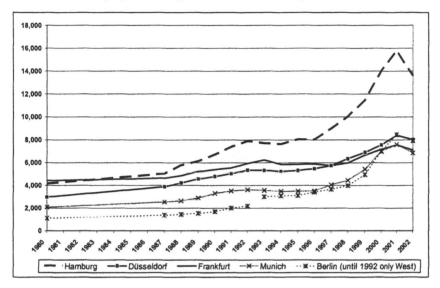

Figure 3.9   Evolution of Advertising Employment in the West German Advertising Centres (1980-2002)

Source: Employment statistics, own calculations

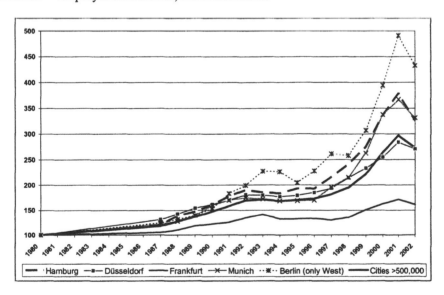

Figure 3.10   Evolution of Advertising Employment in the West German Advertising Centres (1980-2002, 1980=100)

Source: Employment statistics, own calculations

Also the trajectories of Düsseldorf and Frankfurt are remarkable. The former still until 1990 displayed the highest growth rate, from then on yet falling behind and at the turn of the century even developing below the average of all cities bigger than 500,000 inhabitants. Frankfurt, the undisputed national centre of the 'first wave', has continuously lost dynamic in comparison with all other centres meanwhile even ranking almost behind Munich in absolute employment figures.

These differences in the development paths of regional advertising industries are even more clearly reflected in the next Figures 3.11 and 3.12 by means of the differential shift from shift-share analysis, used for the three traditionally most important advertising centres of West Germany (Hamburg, Frankfurt and Düsseldorf) and the newcomers Berlin[21] and Munich, as compared to the absolute change of advertising employment. Since the time series analysis just outlined in the Figures 3.9 and 3.10 has shown different performances of each region in distinct sub-periods we split the overall period under review in four sub-periods which represent the ups and downs of the business cycle. The differential shift is so to speak the difference between the real change rate of advertising employment within one city and a theoretical change rate it would have achieved if it had evolved in the rhythm of all urban centres. It shows thus to what extent the advertising industry in one city follows the rhythm of all cities in a given period.[22]

The figures nicely show the diversity of trajectories of advertising employment in the German urban system. It is particularly striking that Hamburg, besides continuously developing outstandingly, performed particularly well in the two boom periods between 1987 and 1992, and between 1997 and 2001, respectively. Unlike that the two traditional centres, Düsseldorf and Frankfurt, relatively lost in particular in these periods, the former in the first half of the 1980s still revealing the strongest dynamic of all centres, the latter yet constantly evolving below the average. The patterns of the newcomer cities are basically similar; whereas Berlin yet could constantly increase its relative dynamic from sub-period to sub-period, Munich's growth path shows a slight setback during unification.

---

21 In order to guarantee the consistency of the time series we only use the Western Part of Berlin therefore tending to underestimate the relative dynamic at the and of the 1990s given that many of the creative start-ups in the post-unification city took place in the inner city districts of the former East (Krätke and Borst, 2000).

22 For a comprehensive approach to shift-share analysis see Arcelus, 1984, for an introduction Schätzl, 1999.

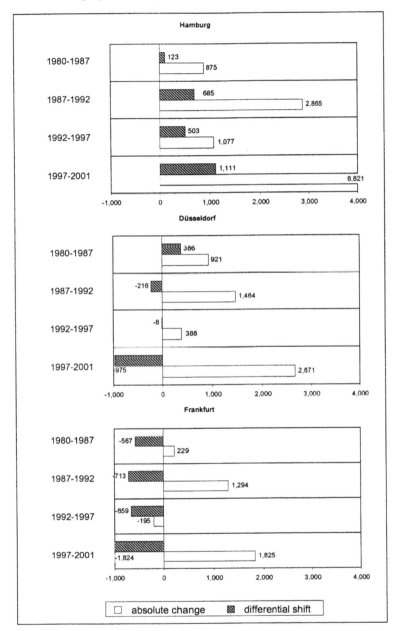

Figure 3.11   Absolute Employment Change and Differential Shift in Advertising – Traditional German Advertising Centres 1980-2001 in sub-periods

Source: Employment statistics, own calculations

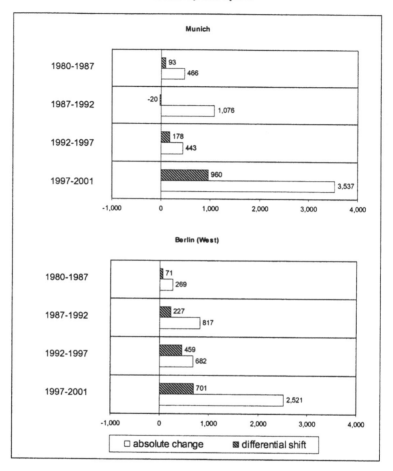

Figure 3.12   Absolute Employment Change and Differential Shift in Advertising – Newcomer German Advertising Centres 1980-2001 in sub-periods

Source: Employment statistics, own calculations

We shall now focus on the internal structure of the advertising industry in each region in order to unravel the eventual specificities which could have been the forces behind the different development paths of the West German advertising regions. Due to data availability we return to the level of metropolitan regions, respectively to their urban cores. In terms of firm size (Table 3.3) it is Hamburg and Rhine-Main which display the highest degree of firm concentration: Only 31 respectively 34 per cent of employment is covered by small enterprises with less than 20 employees, as against 32 and 31 per cent in firms with more than 100 people. The corresponding average values for all metropolitan regions are 40 and 24 per cent, respectively. The Rhine axis, while being the headquarters of the three biggest advertising agencies in Germany, displays an even lower concentration level than the average, with about 23 per cent of employment in firms with more than 100 employees, yet having undergone a process of concentration since 1997 when the same group only amounted to about 14 per cent. The two newcomer regions are nearly equal regarding the dominance of small firms. Berlin, however, also shows an above average share of employment in agencies with more than 100 employees, thereby seemingly reflecting a relocation dynamic towards the capital by big international networks at the very end of the 20th century.[23]

Table 3.3 Percentage of Advertising Employment by Firm Size in Urban Cores of Selected Metropolitan Regions (2001)

	Metropolitan Regions	**Hamburg**	Rhine	Rhine-Main	Berlin	Munich
1-9	25.35	**18.62**	23.05	21.11	29.52	26.89
10-19	14.34	**13.78**	12.45	13.77	13.87	15.48
20-49	20.92	**20.62**	23.66	18.35	15.14	25.63
50-99	14.88	**15.24**	17.54	15.60	15.19	15.96
100-499	20.80	**28.04**	23.29	21.92	26.28	16.05
>499	3.70	**3.71**	0.00	9.26	0.00	0.00

Source: Employment statistics, own calculations

In terms of the internal organization of the agencies, Frankfurt and Hamburg appear to differ substantially. Tables 3.4 and 3.5 allow a view on the occupational structure of the advertising sector, distinguishing 'creative activities' as core tasks of the advertising process from the overhead functions 'management' and 'clerical activities'.[24] According to this criterion Hamburg seems to reveal a distinct type of advertising which despite substantial firm size poses main emphasis on creative work. Rhine-Main, unlike this, clearly shows the main features of big network agencies, with a strong focus on the overhead functions management and office work.

---

23 See figures 4.2 and 4.3 in Chapter 4.
24 Here we use an own classification developed for a report on employment in Hamburg adapted to the specificities of advertising (Läpple and Kempf, 1999, Läpple and Kempf, 2001a).

Table 3.4 Percentage of Advertising Employment by Occupational Category in Urban Cores of Selected Metropolitan Regions (2002)[25]

	Metropolitan Regions	**Hamburg**	Rhine	Rhine-Main	Berlin	Munich
Creative activities	40.10	**45.53**	42.78	42.66	33.10	42.65
Management activities	5.28	**5.34**	5.24	6.23	5.17	6.28
Clerical activities	30.23	**27.06**	30.63	33.56	33.48	29.33
Others	24.39	**22.06**	21.35	17.55	28.25	21.74

Source: Employment statistics, own calculations

Table 3.5 Percentage Change of Creative Activities in Advertising in Urban Cores of Selected Metropolitan Regions (1980-2002)

	1980	1987	1992	1997	2002
Metropolitan Regions	38.51	40.30	41.60	42.09	40.10
Hamburg	**36.43**	**39.29**	**44.29**	**46.61**	**45.53**
Rhine	44.44	44.46	44.08	44.39	42.78
Rhine-Main	39.55	43.87	45.62	42.69	42.66
Berlin	30.72	35.07	34.45	32.12	33.10
Munich	34.83	36.93	41.32	45.33	42.65

Source: Employment statistics, own calculations

The evidence is even more notable when considering the change that has occurred during the last twenty years. Whereas Rhine and Rhine-Main maintain a more or less stable, although above average share of creative activities in advertising, Hamburg has obviously been the most striking structural change. In other words: The growth process of the advertising sector in Hamburg has not just made agencies grow but was able to change the traditional agency structure towards a focus on creative work. This change has constantly occurred since 1980; it led to a distinguishing feature only in the 1990s. The 'new economy' period around the turn of the century has constituted a slight setback which however is likely to be due to the rise of technology oriented internet firms classified as advertising. What

---

25 Unlike in the previous analyses we use the 2002 values here given that the advertising figures for 2001, the peak of the internet boom, are too strongly distorted by internet firms wrongly classified as advertising agencies and thus overemphasizing 'technical occupations', in our data subsumed under clerical activities.

is again surprising is that the internal changes in the Berlin based advertising industries are minimal. Thus, although there has been an enormous rise of new agencies being established there, a focus on creative core work is not yet reflected in the employment structure.

Thus, the shift to the favour of creativity has been and continues to be most evident in the structure of the Hamburg advertising industry and may ultimately have been the vehicle also of its economic success story. This evidence is equally reflected in the creativity rankings collected by specialized journals and sector associations and putting together the different agency's success in the diverse national and international creativity contests the most prominent of which is the International Advertising Festival of Cannes. In a ranking summarizing the years 1990-1998, for instance, five Hamburg based agencies are among the top ten, on position 1, 2, 4, 6 and 7 (Richter and Peus, 1999). In addition to this, a whole body of small enterprises with a remarkable success in terms of both creativity and business performance has risen during the 1990s. The title 'Newcomer-agency of the year' which is awarded annually since 1991, until 2002 five times was won by agencies from Hamburg (Strahlendorf, 1999, p.14; http://www.axelspringer.de/inhalte/pressese/inhalte/presse/passage/40.html, 08/10/2004), and the regional distribution of top 100 agencies founded in the 1990s shows the city's prominence in the sector's recent dynamic as compared to the other advertising centres.[26]

The subsidiaries of the international agencies have largely continued to be based in Düsseldorf and Frankfurt, and they do still constitute – in terms of gross income – the major players on the German advertising market. However, they also have oriented themselves towards Hamburg in manifold ways, partly setting up branches in Hamburg, partly acquiring shares of local agencies in order to exploit the creative potential of the city, partly even shifting their national headquarter northward.

However, the new pattern of the advertising industry in the German space-economy is by no means a finite state. The dynamic development of Munich and, above all, Berlin at the end of the 1990s has shown that the constant competition between the urban centres of Germany continues to reshape the German advertising landscape. Even if the re-established capital, as compared to its actual size, is still on a substantially lower level than the former 'big three' it is very likely to constitute an important hub in the national industry in the near future. We shall see in the next chapter that this scenario already today plays a major role in many of the German agencies' thinking about their future development.

---

26 Ten of the Hamburg based agencies established since 1990 rank among the German top 100 in terms of gross income in 2000. Rhine and Rhine-Main each only account for six according to the same criteria (*Werben und Verkaufen*, 13/2001).

Innovation in Advertising as 'Window of Locational Opportunity': A First Interpretation of the German 'Second Wave'

The fundamental changes the global advertising business has been undergoing in the course of the last about thirty years have substantially affected also Germany in various respects. First, also German advertising has generally adopted a more creative and entertaining style in order to convince a more demanding audience. Second, this increasing need for creativity has strongly been underpinned by a general growth in the importance of advertising, in Germany particularly fostered by an 'exploding' media system since the establishment of private TV channels in the :niddle of the 1980s. The German advertising market has since then experienced a process of 'up-catching modernization' which on the one hand increased the importance of advertising, but on the other hand complicated its success due to the abundance of communication channels.

Third, also the locational pattern, visible in the spatial distribution of advertising employment, has been substantially modified in the course of the ongoing changes. The pattern of urbanness vs. non-urbanness, on the one hand, has been affected most strongly by the general growth of advertising. As professional market communication has come to be a general tool for competitiveness, also firms previously not having used advertising or having made their marketing solely in-house now tend to buy it from external agencies. This mechanism has made advertising relatively grow stronger outside metropolitan regions and outside the biggest cities. However it has thereby not ceased to be a genuinely urban activity, with a continuing absolute dominance of metropolitan regions and particularly their urban cores. What appears more interesting and more closely linked to the qualitative changes inherent to the 'second wave' of advertising is the shift in the employment distribution among the twelve metropolitan regions and particularly among the three leading advertising centres, Hamburg, Frankfurt and Düsseldorf. The restructuring of the German advertising industry happened definitely to the detriment of Frankfurt on Main, and, since the beginning of the 1990s also to the detriment of Düsseldorf. Hamburg has been able to establish itself as the creative capital of German advertising, the agencies in the North-German city-state most successfully performing on national and international creativity contests. The 'new economy' boom at the end of the 20th century has shown that seemingly a next round of restructuring has begun whose final results however cannot be assessed yet.

Fourth, also the increasing splitting up of the agency landscape in global mega communication holdings, on the one hand, and small and fast-growing creative 'hot shops', on the other, is mirrored on the German market even if employment data does not reflect this unequivocally.[27] The 'globalizing' tendency is visible

---

27 Employment data on the one hand tends to underestimate concentration due to the structure of the global groups, which appear as separate firms, and, on the other, to underestimate dispersion due to the missing assessment of self-employed persons.

both in the acquisition of local agencies by global networks either as growth strategy or as a means to exploit creative potential and in the reflections of mergers on the global level on the reorganisation of national subsidiaries. The decentralizing tendency is visible both in the increasing number of freelancers in the sector[28] and in the whole body of successfully established agencies throughout the country from the late seventies onwards. Hamburg's rise from 1980 ultimately concurred with the success stories of two newly founded key agencies, Springer & Jacoby and Scholz and Friends, established in 1979 and 1982 and reaching the top 20 of agencies on the German market in terms of gross income in 1989 and 1986, respectively. Together with one spin-off of Springer & Jacoby, Jung von Matt, which was set up in 1991, being already in 2000 the 19th biggest agency in Germany, they are the protagonists of creative German advertising.

This creative turn was also associated with a substantial reorganization and decentralization of the formerly hierarchic and vertically integrated structure of the advertising agency, through the creation of small and largely independent units which allowed a close co-operation of account managers and creatives and which flattened the hierarchies of the firm, giving a large degree of responsibility to employees in very early stages of their professional careers. This in turn fostered spin-offs from the parent agencies through which gradually a cluster of new agencies emerged. These agencies established from the late 1970s until today altogether represent the German version of the 'second wave of advertising'.

Thus, the alteration of German advertising through the rise of the 'second wave' consisted in the interaction of different changes, regarding style, general importance, spatial and organizational structure of the advertising industry. Summarizing the overall process one can argue that these changes both presupposed and implied fundamental innovations in terms of product, process and organization. The new product was the 'new type of advert' (Lash and Urry, 1994, p.140); this was achieved by a modification of the process, bringing closer together the two core functions account management and campaign development through the introduction of account planning (ibid., p.141). This in turn took place in a new 'hybrid' and decentralized type of advertising agency in which both the close interaction of 'client liaison function' and 'creative development' could unfold and through which motivated personnel was enabled to quickly assume responsibility for the very performance of the agency.[29]

This closely inter-linked set of innovations took place outside the pre-existing structures of the first wave, in terms of both agencies and locations. That is to say, the organizational change did not occur as a restructuring of the vertically

---

28 The share of self-employed persons without employees in the total advertising employment increased in West Germany from 1991 to 1997 from 21 per cent to 26 per cent, in the cities with more than 500,000 inhabitants from 22.5 per cent to 27 per cent in the same period (Microcensus scientific use file, own calculations).

29 Lash and Urry (1994, p.140f.) only mention product and process innovation. For the understanding of the German situation, it is yet mainly this organizational change which appears crucial for the dynamic occurred in Hamburg. See the following Chapter 4.

integrated American type full-service agency of the first wave, but through the establishment of new enterprises, and it did not start from the dominant centres of the former period but produced a new centre of the German advertising industry. Reflecting these findings in the face of the theoretical arguments on spatial restructuring specified in Chapter 2, the period in which the 'second wave' of advertising was coming into existence can be conceived of as a 'window of locational opportunity' (Storper and Walker, 1989, p.75), that is, as a period in which the main drivers of innovation are on the one hand relatively free as to their location and, on the other hand, tend to locate outside existing centres in order to evade the rigidities of the formerly hegemonic structures. The predominance of old type advertising agencies particularly in Frankfurt but also in Düsseldorf appears to have made both cities little apt to the fundamental change so that it had to be triggered elsewhere.

As put above, Storper and Walker's argument had originally been launched to explain the rise of the Silicon Valley as a non-urban high-technology growth pole that emerged so to speak in a profoundly rural environment. They held that the region was completely 'produced' by the upcoming industries and their 'factor-creating and factor-attracting power' (ibid.). In the case of German advertising the situation is different since the rising industry did not evade cities in general but produced a new advertising cluster in an existing urban core in which a basic critical mass of the industry and its infrastructure already existed. Thus, the locational shift of German advertising in the course of the emerging 'second wave' constituted a specific evolution in two respects: On the one hand, it underpinned the role of advertising as a truly urban industry; on the other hand it signified a typical phenomenon for the de-central German urban system which can be read only to a limited extent in the hierarchic categories of urban/non-urban. Thus, unlike the spatial restructuring processes triggered off by the rise of high-technology industries the restructuring of advertising did not challenge cities as production site in general but appears to have exploited urban space in a different way.

But how can this new way of using cities as a production site for advertising be characterized? How do new production structures and urban structures interact across urban space? And how have the 'leading firms in the rising ...' new advertising '... industry' (ibid., p.74) unfolded and do still unfold their factor-creating and factor attracting power? How do they thus 'produce regions' in a given both urban and national industry environment? In the following chapter we shall try to get deeper into the logic of the 'second wave' of advertising in Germany by, first, looking closer at the process of innovation having driven by the new advertising players and, second, by examining how this innovation impinged on and changed the national agency landscape. Thereby we aim at finding out not only how the action frameworks of different agencies have changed but also how this change interacts with the social networks and spatial environments in which these agencies operate.

Chapter 4

The Firm Perspective: Reading the Restructuring of German Advertising as an Innovation Process

Agencies as Drivers of Change: Entrepreneurship and Organization

As seen above, it was basically an innovation process driven by a group of newly established agencies which lay at the heart of the changing territorial pattern of the German advertising industry. Through developing a new advertising style underpinned by new ways of process and firm organization these agencies succeeded in establishing an alternative to the traditional vertically integrated American advertising firm, thus emerging as serious new players in the German agency landscape, thereby ultimately propelling its reorganization.

Put conceptually this process involved two key notions of dealing with economic activity which have influenced the discussions in many subparts of economically oriented social science in a significant but also contradictory way. On the one hand this is 'entrepreneurship' which gained in importance mainly through Schumpeter's work and his definition of the concept as 'doing things that are not generally done in the ordinary course of business routine' (Schumpeter 1988b (1949)). Entrepreneurship thus is a matter of the new and unexpected, of the breaking out of pre-given frameworks of action. Logically many of the classic accounts on Silicon Valley and other regional success stories already discussed in Chapter 2 (e.g. Storper and Walker, 1989; Saxenian, 1994a) explicitly stress the personality of the entrepreneur as key actor for the success, but also regarding its continuous innovativeness and economic prosperity. On the other hand it is the firm as one particular form of the 'organization' of economic activities which, beginning with the scientification of management in Taylorism and with Coase's work on transaction cost, and continuing in the accounts on 'economies of scale and scope' in Fordist mass producing corporations (Chandler, 1977, 1992; Piore and Sabel, 1984), has drawn the attention precisely to the 'routines' and the continuity facilitating the coordination of production processes and guaranteeing the return on investment. Yet, the firm as an organizational entity has not only been discussed as a way of optimizing and rationalizing production; processes of 'lock in' identified for regional institutional settings[1] also apply to organizational routines and specific firm cultures, thereby being likely to act as barriers to change in periods of fundamental transformation (Schoenberger, 1997, 2000).

Thus, the dualism of entrepreneur and firm, as regards their role for economic competitiveness and success, reflects a similar kind of dialectic as discussed for the relation between innovative actors and their environment[2] and as visible in Florida's (2002) confrontation of 'creativity vs. organization'. In theoretical terms it exemplifies Giddens' (1984) 'duality of structure': Structures – in our case the organizational pattern of the firm – are necessary conditions of individual action, at the same time however acting as its constraints. Entrepreneurship is thus one way of individual action aiming to deal with these constraints which, yet, in turn needs structures to function at all.

The aim of this chapter is to examine the process of change German advertising has undergone in the course of approximately the last 20 years from the micro-perspective of advertising agencies, respectively of full-service agencies which constitute both the major part of the industry and the pivotal point in the advertising value chain.[3] In this context the duality of entrepreneurship and organization is taken up and further developed in a multidimensional way. Two aspects seem worth being anticipated: First, while the innovation process was in fact based on entrepreneurship, organizational features regarding the work process, the firm structures, the patterns of collaboration etc. were inherent part of its 'substance'. Interestingly, the organizational dimension of innovation aimed precisely at fostering entrepreneurship within the context of organization. Second, the innovation obviously took place within and as an alternative to a given agency landscape which was, on the one hand, influenced and ultimately changed by the innovation and its organizational features. On the other, the traditional agencies cannot only been considered as being affected by the innovative players. They definitely had to react and they did it, again also in terms of their organization, this altogether definitely changing the pattern of competition in the national advertising market.

The dramaturgy of this chapter basically follows these two perspectives of change. It starts with the story of one of the two key players of the Hamburg success story, Springer & Jacoby, which with its establishment and its accentuated firm culture most radically represented a new way of advertising in the German market and which thereby can be most clearly be characterized as the key agent of innovation. In a second step we shall try to highlight how the changes driven by the key players interact with the different groups of agencies characterizing the

---

1   See pp.16ff. in Chapter 2.
2   Ibid.
3   According to the firm database of the Hamburg chamber of commerce, 55 per cent of the 6,173 enterprises in the city's advertising sector were classic agencies – as of July 2001 (www.hamburg.ihk.de/produktmarken/produktmarken.htm, 10/09/2001). See also pp.48ff. in Chapter 3.

German advertising both influencing them and being in turn influenced by their reactions. Whereas this step discusses each type of agency separately, the chapter's final section will attempt to provide a comprehensive interpretation of the industry's restructuring, figuring out the basic logic behind the process of change as well as the dialectic of segmentation and interdependency within the overall agency system, and the way how both interact with the spatial context it is interwoven with.

The Substance of Innovation: Exploring the Springer & Jacoby Story

Springer & Jacoby (S&J) was founded in October 1979 by Reinhard Springer, a trained advertising professional who – after several years of work in different agencies throughout the country – simply wanted to start his own business in his hometown Hamburg. The situation of advertising in the city at that time was not considered to be very favourable:

> ... and the colleagues said that I was stupid since Hamburg was simply no place for advertising, [...] because here was the 'mother' LINTAS, that was at that time the biggest agency in Germany, it was regarded somehow as university of marketing and classic advertising, but besides it Hamburg was advertising backwater (personal interview, 2000).

His partner Konstantin Jacoby, a copywriter, only four years later joined the agency. Springer and Jacoby had previously been colleagues at the German headquarter of the Swiss agency GGK in Düsseldorf. To a certain extent the GGK constituted the creative pioneer of German advertising in the 1970s.

Thus, creative 'second wave' advertising had already had antecedents, its roots reaching back to the 1950s and 1960s when the first creative professionals began to set up their own agencies (personal interview, 2000). The agency 'Team', maybe the archetype of a successful national agency foundation, originated from an art department established in 1953 by two just graduated commercial artists (Merkel, 1988). In 1972 it was acquired by the American network BBDO using 'Team' as gateway to the German market. In the 1970s its role was taken by the GGK. Both 'Team' and GGK were based in Düsseldorf thereby constituting so to speak a more artistic counter pole to Frankfurt (personal interview, 2000).

Hamburg's rise as creative capital of German advertising to a large extent originated from these two agencies, not only using them as 'role models' but really being produced by them: Scholz & Friends was a spin-off of the Hamburg office of Team/BBDO established even by one of the original 'Team' founders, Jürgen Scholz (see Scholz, 1998) and Springer & Jacoby was founded by two former members of GGK.

Despite this actual similarity, we concentrate our following account of the innovation process on Springer & Jacoby alone since the two key players substantially differ in how radically they stand for a 'new' orientation in advertising. This difference was already mirrored in the nature of each agency foundation. Scholz & Friends (S&F) already started as a relatively big player with important Hamburg based clients 'inherited' from 'Team/BBDO'[4] whereas Springer & Jacoby (S&J) was established without such a guaranteed financial base. Thereby they also represented two more generally distinct models of agency start-ups, as we shall point out later in more detail (p.103). Moreover the explicit focus on creativity was still more accentuated in the case of the latter.

In the dreary monotony of German advertising the Hamburg based agency had from the very beginning consequently banked on entertainment as marketing tool (Boldt, 1996, p.106).

Thus S&J most paradigmatically represents the very innovation in German advertising in terms of *product*, *process* and *organization* pointed out above.[5]

About three months after its foundation the agency succeeded in acquiring its first client, a Hamburg based shoe retail chain, today with stores in overall Germany. The idea to both develop an encompassing image campaign for the chain instead of just inserting ads, and to underpin it with surprising communication ideas was at that time a substantial innovation for retail advertising and it brought sustainable success for the client, still today being the only German shoe retailer constituting a brand on its own. With this campaign Springer & Jacoby laid the foundations of its further success. Its rise to the most prominent agency in Germany, however, began only after the entrance of Konstantin Jacoby. From ten employees in 1983, Springer & Jacoby grew to 35 in 1984, 72 in 1987, 120 in 1989, more than 300 in 1991 to 518 in 2000.[6] In 1984, the first international advertising award at the Cannes festival was won (Springer & Jacoby, 2000); the amount of national and international awards obtained between 1984 and 1989 was 130. Springer & Jacoby was clearly dominating the creative realm in Germany during the 1980s.

That this success in terms of creativity could be transformed into a real *product* innovation, that is, it was also accompanied by an outstanding business performance, was mainly fostered by two factors: First, the introduction of private TV in Germany from 1984 onwards supported the agency to unfold its capacity in making unusual TV commercials. In the first years of its existence S&J had very strongly focused on cinema commercials and had succeeded – mainly based on Konstantin Jacoby's talent in script writing – to completely change the image of this type of advertising:

---

4   In the case of Scholz & Friends the rise of Hamburg was thus substantially client-induced. Until today their growth has been closely linked to big players of the city's food and tobacco industry.

5   See p.77 in Chapter 3.

6   Figures from *Frankfurter Allgemeine Zeitung*, 25/03/87, *Die Welt*, 02/02/89, Springer & Jacoby, 2000 and http://www.sj.com/german/facts/daten/inhalt.html, 10/09/2001.

... and we made films that enticed people to go to the cinema. Well, normally people begin to leave the cinema when commercials are on, [...] and we were actually the first, who – in 1983, 84, 85 – who made films that swept the audience off their feet (personal interview, 2000).

Private TV provided the possibility to apply this conception of an entertaining advertising film on a larger scale. S&J even hoped that the widened space for advertising in television would improve the general quality of TV-spots since corporate executives had now the opportunity to experience the perception of advertising after 8 p.m. at home in their families.[7] Still today TV and cinema advertising accounts for 42 per cent of the agency's activities, thereby covering by far the largest share in S&J's media mix.[8]

The second factor, which substantially fostered the growth of the agency, was that it succeeded to 'sell' its unusual advertising to big clients. The first 'blue chip' on Springer & Jacoby's client list was the German household technology corporation Miele in 1985. However the agency's final breakthrough occurred when winning the entire Mercedes-Benz account in 1989.

... and we have always been so ambitious to say: 'We want fantastic clients, we look for real brands' and, as Konstantin Jacoby is a car fan, it was my idea to try a car brand and we tried Mercedes. And so we acquired for seven years, called them again and again during seven years and asked: 'What do we have to do to become a possible partner for you?' And they did of course find every possible condition and we did not stop and continued to structure ourselves in the required direction, so that we one day were able to say: 'Now we can do what you want'. And at that moment we were lucky that Mercedes was experiencing a low regarding its image, and consequently also a low in sales. And when a client is doing badly, he tends to get courageous, and so Mercedes said: 'Well, why shouldn't these odd boys there in Hamburg do the job. Our image cannot become worse than it is now. So they'll get it.' And within two or three years we succeeded to turn the image of Mercedes so that they enormously got wind behind them. And they are still today grateful for that and on the occasion of decennial collaboration they made a big party (personal interview, 2000).

The agency's success in winning 'blue chip' accounts was thus the result of an enormous ambitiousness and obstinacy on the part of the founders (Glabus, 1991). Moreover, it is striking that they benefited precisely from the gradual sector change inherent to the 'second wave' on the client side, above all from the rise of the car market that during the 1980s became the top advertiser in Germany. Car advertising had always been important but firstly not for luxury cars, secondly rarely focusing on TV advertising, and thirdly not dominating the post World War

---

7 From *Die Welt*, 02/02/89: 'Durch die heißeste Agentur weht ein steter Hauch von Frischluft'.

8 From http://www.sj.com/german/facts/daten/inhalt.html, 10/09/2001.

II media world in the same way as food, tobacco, detergents etc. did.[9] These 'classic' brand articles are still today underrepresented in the client structure of the agency.

Hence it was above all two important changes on the market side, that is, an altering media system with TV as new dominant communication channel, and a changing client structure, as well as importantly the agency's capacity to deal with them, that helped S&J to transform creative or entertaining advertising into a business success, respectively to introduce a substantial innovation in terms of the *advertising product*. Yet, what appears even more important in terms of its overall impact on German advertising, were two closely linked innovations on the supply side, in terms of *process* and *organization*. Put in other words: Springer & Jacoby's main innovations lay behind the product, respectively in the way how the agency managed to act as a basis on which creativity could and still can unfold.

As pointed out above, the crucial *process* innovation of 'second wave' advertising was considered to be the introduction of account planning as a means to reconcile creative development and account management (Lash and Urry, 1994, p.141). In the case of Springer & Jacoby the reconciliation remains important however less by means of a 'third column' in campaign development, but rather based on a common enterprise culture. In the founder's words:

> ... I am a somewhat atypical entrepreneur, thus it was not my main objective to become rich or to make money. My main objective was rather to have a community, that is, to make something unusual together with some people I like. [...] and we never considered ourselves as a 'normal' enterprise but [...] rather as a community. And, based on this idea, we always saw ourselves as a school, in a double sense. For, first, we have continuously learned until the present day and, second, we were a school in that we trained people (personal interview, 2000).

The key concepts, through which this particular understanding of an advertising agency is expressed, are 'community' and 'school'. That is to say, besides uniting the creative and the business part of advertising through personal and friendship ties and through developing a common – internally coherent but externally distinctive – understanding or even philosophy of advertising S&J considered itself as a collective learning system, again with two dimensions: On the one hand, it encouraged individuals to adopt the conventions or patterns of behaviour within the community. People were trained to become part of the agency's distinctive culture.

> We said: 'We are unique, and as we are unique [...] we have to train our own people.' And since, as a community, we developed an own culture, with a basic law, certain

---

9   In the 1980s tobacco, TV sets, spirits and retail trade were the least dynamic economic 'sectors' in the German market regarding advertising expenditure. Computers, general image advertising and the mass media showed the highest growth rates. The car market ranked at no. 8 in terms of its percentage change (ZAW, 1990).

convictions, rules and so on, we in fact had to train our people on our own. We did of course do that; and it is an enormous effort (personal interview, 2000).

The objectives and principles of this learning process were institutionalized, being laid down in a basic law developed and written down by the heads and containing the basic values and principles of Springer & Jacoby. To achieve an outstanding advertising which should be 'simple, imaginative and exact'[10] a fair and human co-operation was considered to be a crucial pre-condition (Boldt, 1996). This fairness was articulated in a series of individual rights each member of the community had: a right to be promoted, a right to well-being, a right to make mistakes, to contradict, to be protected, to transparency etc. In addition the institutionalization is also reflected in the daily work of the firm: The management is regularly assessed by an agency 'culture-check' in which the employees evaluate the executives' work according to the basic principles of the agency (ibid.), and there are annual target agreements for every employee. That is to say, besides the idea of the agency as a coherent entity it is the orientation of the firm along the individual member of this community, respectively to his or her capabilities in order to exploit the potentials of the human capital as much as possible.

On the other hand, once the community is established learning means its ability to cope with the changing requirements of the business, above all on the part of the clients. The way how the acquisition of Daimler-Benz was worked for displays how the agency functioned as a learning community. Also the results of these learning processes are evaluated: by the clients who annually assess the agency's work. The overall performance of the firm is measured by means of the four criteria client, culture, creation and cash (ibid.).[11]

The focus on the individual's capabilities does also play a major role in this second dimension of learning. The overall learning capacity of a community is enhanced through exploiting the capacity of each member as fully as possible. And this principle is reflected by S&J's *organizational* innovation, that is, to let the agency grow in a decentralized structure of single, legally independent units:

And so we had lots of very ambitious young people here [...]. And since we had the character of a school – in normal firms higher ranks in the hierarchy always tend to keep the lower ones low in order to have a calm life – we didn't do that this way. Instead we wrote down a sort of basic law, and it said: 'In this agency everybody can get as far as he or she wants to, and as fast as he or she wants to.' [...] There was also no reason to be afraid, since when there was a young and dynamic talented guy, he could open up a new unit, and you can continue this as much as you like (personal interview, 2000).

---

10 In the (German) words of the basic law the advertising of S&J should follow 'the three Es' (*einfach, einfallsreich, exakt*). See Boldt, 1996, p.112.

11 In German the criteria are called 'the four Ks: *Kunde, Kultur, Kreation* und *Kasse*' (Kemper et al., 1999).

The unit-principle enabled Springer & Jacoby to maintain both the community idea and the focus on individuality although the agency kept growing. Small units facilitated strong personal ties and a close co-operation between different agents within campaign development. At the same time individual employees were not only trained in a classic sense but were encouraged to assume an entrepreneurial attitude, thus feeling personally responsible for the performance of the firm. As a consequence S&J already in 1987 consisted of three agencies (with 72 employees), in 1989 of six (with 120), in 1996 of 11 (with 240), six of them for classic advertising and further five units for special tasks such as administration, media buying, sales literature, lithography, and new media, and in 2001 of 18 legally independent firms, four of them outside Germany.[12] Thereby they combined the idea of easy labour promotion and entrepreneurship with the possibility of a big firm to achieve economies of scale in terms of facilities, services and specialized activities.

The strategy of both giving a strong importance to the individual capability of each employee and providing them with opportunities to advance was carried to an extreme when the founders sold 50 per cent of the agency to 38 members of staff in 1994 for a cut rate in order to let them benefit from the overall profit. With this step the agency developed to a certain extent a small and 'democratic' version of the globalizing tendency within the 'second wave', that is, the submission of the agency business to the logic of financial holdings. The founders themselves represented this financial part of innovation more strongly. At the peak of their involvement in advertising activities in 1996 they had shares in 23 firms, four of which were mere holdings to manage the participations (Boldt, 1996). Nowadays this situation has changed given that a minority share of the Springer & Jacoby group is owned by one of the top three advertising organizations in the world, the Interpublic Group of Companies. In the course of this internationalization the structure of S&J participations was reorganized. As regards the core agency the involvement of the labour force in a participation strategy however continues to exist, combining the idea of a community with the principle of individual promotion. 'Everybody can get as far as he wants to' in this sense means that the agency offers the possibility to its staff to fully become entrepreneurs without leaving the firm.

Put more conceptually, through their overall strategy Springer and Jacoby succeeded in mobilizing and speeding up the agency's internal labour market by broadening the upper end of the firm hierarchy thereby widening the bottleneck that hinders promotion of talented and ambitious employees. However, this strategy did not remain without repercussions on the external labour market since, first, even a widened bottleneck does not always allow for every ambitious person to advance within the firm, and, second, to be a profit-sharing head of a unit is not enough for everybody's ambitions. That is to say, encouraging the ambitions of the

---

12 Figures from *Frankfurter Allgemeine Zeitung*, 25/03/1987; *Die Welt*, 02/02/1989; Boldt, 1996; http://www.sj.com/german/facts/daten/inhalt.html, 10/09/2001.

own labour force automatically resulted in an enhancement of both labour mobility[13] and new firm start-ups not only by 'real' S&J-people: 'Dozens of agencies […] were set up throughout Germany, with one single purpose: to make an advertising as intelligent as that of Springer & Jacoby' (Boldt, 2001, p.172). In this sense, the idea of an S&J-school of unusual, entertaining advertising influenced the sector in the whole country, in a very practical sense:

> Well the whole quality in advertising has of course been improved, since when you have some draws in front, they tend to inspire all the others. And lots of things are copied, you can see it or at least you notice that it has been inspired. In addition to this, from the Springer & Jacoby-school, have a look around, you'll find in nearly all more or less renowned agencies in Germany Springer & Jacoby-people. That has – even if homoeopathically thinned down – of course always some effect, that is logical, isn't it? Well, as a school we did certainly help to raise the national level (personal interview, 2000).

Nevertheless the relation between S&J-community and the overall sector is still strongly shaped by the distinctiveness of the new school:

> We have trained all our people on our own, and up to the present day Springer & Jacoby is still not completely compatible. When people come from absolutely different agencies, […] they have extreme difficulties and nearly always fail. Thus, who succeeds to advance here are either own young people or those who come from the 'breakaways', thus from Jung von Matt, since they began with 15 S&J-people and were actually a Springer & Jacoby unit. And still today they have extreme similarities (personal interview, 2000).

That is to say, the innovations in *product*, *process* and *organization* inherent to the business strategy of Springer & Jacoby on the one hand fostered the rise of the agency itself to become one of the TOP 10 on the German market. On the other hand, the processes of individual and collective learning originally having started within the agency were transferred to a supra-agency level thereby influencing the whole structure of the advertising industry in Germany and particularly in Hamburg. This occurred mainly 'through skilled labour mobility within the labour market' (Camagni, 1991b, p.130), both between agencies and as firm spin-offs, as a side effect of the individual motivation, and through 'imitation processes' (ibid.). Thus innovation in the German advertising sector, at least as regards the collective learning processes, strongly resembles the mechanisms shown by GREMI researchers for the high technology milieus of the 1970s and 1980s (ibid.), however not unanimously including the local delimitation.

---

13 On the other hand the unit structure also facilitated to keep the necessary mobility of labour within the firm. Thereby it also enabled the agency to deal with the frequent changes in work volumes without using too many freelancers usually more expensive than in-house staff. The average number of freelancers is only 35 per year, that is, less than 10 per cent of the permanent staff (personal interview, 2000).

Yet, despite the fact that the Springer & Jacoby-school has had substantial repercussions within the whole sector, the agency's new style, enterprise culture and enterprise structure still continues to constitute an own distinctive way of doing advertising. Its overall impact on the overall agency pattern (not only) in Hamburg thus has to be read as a complex relation between continuity and change.

The Impact of Innovation: Continuity and Change in the German Agency Landscape

The change in the German advertising landscape triggered with the rise of the new pioneer agencies of the 'second wave' can be read on two levels: First, simply put, new players have appeared, changing the global-local dichotomy based on differences in the size of the market agencies cater to, previously discriminating between the agencies, to a more complex pattern of 'global vs. local' and 'old vs. new'. Figure 4.1 both serves as illustration of this new pattern and depicts the agency typology the following discussion is based on.

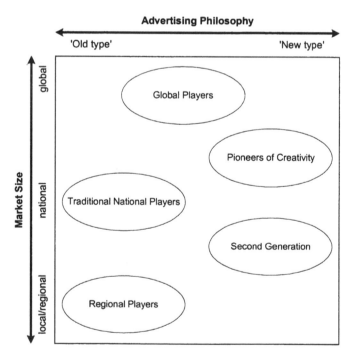

Figure 4.1   The Pattern of Advertising Agencies in the German 'Second Wave'

Source: Own illustration

Five agency classes can be identified, two groups of 'new type' agencies with the two 'pioneers' of the 'second wave' in Germany and their 'second generation', emerged as the 'side-effect' of the pioneers' internal innovation processes, and three groups of traditional advertising firms differing by the size of their market area: the 'global players' that is, the international network agencies, the 'traditional national players' mainly set up in the first decades after World War II, and the small and locally oriented 'regional players' characterized by their strong dependence on a regional client base.[14]

On a second level, the framework of action for each player has altered. The conflicting tendencies of massively increasing dependence on global financial mechanisms, on the one hand, and the emergence of a more entertaining advertising style covered by new and innovative players – according to a lot of commentators – has resulted in a 'disappearance of the middle' (Leslie, 1997, p.1023), that is, in a squeeze between centralizing and dispersing dynamics particularly the traditional middle range of agencies is subject to. As nearly always, also here reality appears to be more complex given that all agency groups are confronted with the transformed and still transforming conditions of their business. Therefore we shall now try to examine their different market positions in a comprehensive way, considering both their traditional backgrounds and the challenges they are presently facing, as separate, segmented frameworks of action in which the different players operate, thereby trying to work out the common basic mechanisms as well as the variations between them.

In order to focus on the main 'arena' of change the analysis will start from and emphasize the most 'active' forces in the agency landscape. That is to say, we shall begin with the three groups next to the top right-hand corner of Figure 4.1, the 'global players', the 'pioneers of creativity' and the second generation from there proceeding to the rather 'reactive' groups, the 'traditional national players' and the 'regional players' which both can rather be considered as being affected by change then as being inherent part of it. In addition, the concentration on the main competition between the traditionally dominant players and the main carriers of innovation will be reflected in different degrees of thoroughness between the sections.

---

14 Of course there is always a certain degree of fuzziness in such a categorization. It omits processes of change firms are undergoing in the course of their lifecycles as well as the interdependencies between and internal differences within the groups. From our perspective they however serve to show the basic mechanisms at work in the contemporary advertising industry as well as the different standpoints in this context.

The Global Players: The Ambivalence of Global Advertising and the Struggle for Creative Reputation

As discussed above, the supremacy of global advertising networks in German advertising is still evident and there is no indication that this pattern may basically change, at least over the medium term.[15] Yet it would be misleading to derive a unilateral dominance of global players on the German market from this evidence. As the agency pattern has changed and diversified both globally and nationally and as operating on different national markets is an increasingly complex task, involving various strategies of culturally embedding global messages in local markets, traditional global players are generally confronted with growing competition. In very general terms, their position in this context is determined by two key aspects, that is, a 'normal' *ambivalence of global advertising* in terms of its performance in national markets and the *specific challenges* coming from the new creative players.

To begin with the first: Agencies which are part of an international advertising network are, on the one hand, indeed part of a global communication process, carrying out single parts within big advertising campaigns for big globally operating clients. This very fact guarantees a high degree of both professionalism and reputation. This is all the more valid since global communication groups provide a series of additional resources and infrastructures such as world-wide market research or specialized agencies for the different communication channels. Their advertising approach is thus 'naturally' global and integrated. The following quotation of a local office manager gives an idea of such an approach and its reflections in the work process:

> We are one of the few agencies world-wide that do not only call themselves international but that also work internationally. Oh, when was it? In January, Malmö called us: 'Do you have anybody who can work on coffee? Two people have just left us.' We checked it and: 'Yes, we have one.' So he went. Yes, we do that sometimes with Malmö and at the moment an English colleague is working here. So, you can interlink it. [...] When we had the world-wide XXX account there were not the New York people making the campaign and saying to the rest of the world: 'Just do it as we have told you!' but there was a team sitting in New York, consisting of eleven nations. Eleven people from eleven countries (personal interview, 2000).

On the other hand, within a global integrated approach a single branch office tends to have a rather passive role. This is, first, based on the fact that much of its work suffers from the so-named 'not-invented-here-syndrome' (personal interview, 2000), which stands for the boredom of adapting the creative development of others – even if done in a globally assembled team – to local conditions in terms of languages, media formats etc. Second, and probably more

---

15 See Table 3.2 in Chapter 3, p.60.

importantly, its room for manoeuvre is limited since a lot of strategic decisions are made outside the agency, however with significant impact on the local level:

> The big global consumer groups all have several regular agency networks for their whole bunch of products. And from time to time they let those agencies pitch for a particular global account, so to say to 'refresh' the campaign. Recently the Frank Mars group did this for one of its chocolate bars. And we lost the global account. As local agency, it is by no means your fault. You even cannot influence it. Nevertheless, at the end of the day you lack the money (personal interview, 2000).

This is all the more serious given that the local agency is subject to the targets set by the global headquarter and also has to contribute to its overhead. In addition the global group restricts local strategic action through its very client structure, given the 'conflict clause'[16] which can even hinder own efforts of cross-border activity:

> Some years ago we held the account of one of the new private suppliers of telecommunication services. It wanted to widen its activity successively throughout Europe and asked us to provide support in market communication. In Spain however the network was working for the national monopolist. Consequently we were not only unable to get the account for Spain, but in the end we lost the client totally (personal interview, 2000).

The specific challenges imposed by the 'creative turn' have tended to reinforce this ambivalence, in fact with particular repercussions on the labour market of creative professionals. It seems evident that both the 'not-invented-here-syndrome', the missing scope for strategic action in a local agency and the general difficulties of working in a big multi-unit organization, in comparison with Springer & Jacoby's 'everybody can get as far as he or she wants to', make the local affiliates of international advertising networks relatively unattractive employers. A management member of a young and fast growing agency describes this nicely:

> Just to give an example: a colleague of mine, he was working in an international agency, on a margarine account, Lätta or Rama or something like this. It took them two years to create a strategy. I mean, that's going beyond a joke. Two years for a margarine, that's ridiculous, isn't it? [...] In fact, people complain that the work in such agencies is just policy. It is by no means purposeful work, no fight for a common project, but just for power, influence etc. (personal interview, 2000).

Of course, there are possibilities to compensate this lack of attractiveness by offering higher salaries given the larger financial resources of a global holding. However, the professional scene in general perceives this as a sort of 'damages' (personal interview, 2000).

---

16 See Chapter 3, p.50.

There are basically two extreme variants to deal with the 'global vs. local'-dilemma necessarily characterizing their business and having been accentuated during the sector's transformations of the last 25 years (Ziegler, 1994): an explicitly 'global' strategy banking consequently on the assets of a global network and its strong client base, and a 'multinational' strategy trying to strengthen the 'local content' of the business, respectively to intensify the local agency's involvement in the national market. They can of course never be found in a pure form and how networks opt to develop is not only a matter of how they see themselves affected by the recent challenges but mainly depends on the specific firm history. Whereas, for instance, the centralized 'global' approach is mainly followed by the early movers of the 'first wave', J. Walter Thompson and McCann-Erickson as well as by other agencies originally linked to one big client, such as the Marlboro-agency Leo Burnett, an orientation to the 'local content' can be found in the case of most of the American agencies internationalizing after World War II, like BBDO, DDB and Grey. They internationalized mainly by buying strong local agencies thereby equally acquiring their local client base. To avoid conflicts partly only minority shares were bought out so that the local agency largely maintained its independence, yet at the same time both widening the spectrum of the global player's activities and improving its results. In Germany the national market leader BBDO constitutes the ideal type of such a 'local content'-strategy with an 82 per cent share of German clients and an ownership of the national group distributed among 100 partners.[17]

The difference between the different approaches is not necessarily reflected in the client structure given that there are – besides the organization of the network – further factors shaping the 'downstream-embeddedness' of the national subsidiaries, like the individual performance of the branch office, its success in pitches, but also the reputation and personal involvement of the management staff in the national scene. For instance, McCann-Erickson, in 2000 succeeded in winning the accounts of Lufthansa and Deutsche Bank given that it was able to offer a full service for their global business.[18]

However, taking BBDO and McCann-Erickson, respectively, as paradigmatic cases for each strategy, the differences are obvious. Figures 4.2 and 4.3 contrast the structure of locations of both groups in the German territorial system, distinguishing between different specializations and different forms of relations between parent company and branch offices: Whereas the geographical pattern of offices is basically the same – apart from the different headquarter cities Düsseldorf and Frankfurt – with two strong additional clusters in Hamburg and Berlin and some disperse offices due to specific clients, above all the patterns of relations differ, with a McCann-Erickson structure seemingly steered from the national headquarter in a unilateral way and a more 'fuzzy' pattern in the case of BBDO, by the majority consisting of minority participations and 'parallel' networks.

---

17 Figures from *Frankfurter Allgemeine Zeitung,* 5/10/2001.
18 From http://www.mccann.de/kunden.html, 18/12/2001.

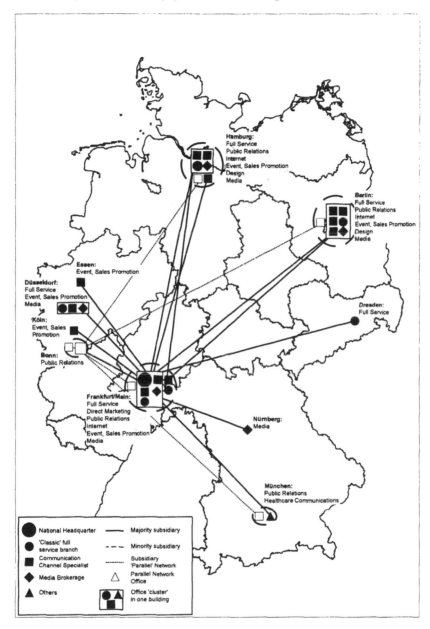

Figure 4.2   The Structure of Locations in the German McCann-Erickson Group

Sources: http://www.mccann.de; http://www.gwa.de; cartography Frauke Funk, Andreas Beekmann, last update 20/02/2002

Creativity and Space

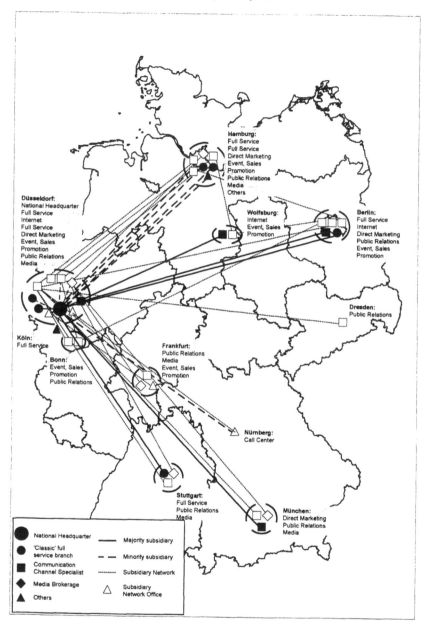

Figure 4.3   The Structure of Locations in the German BBDO Group

Sources: http://www.bbdo.de; http://www.gwa.de; cartography Frauke Funk,
Andreas Beekmann, last update 20/02/2002

One interesting aspect, less related with the global players themselves but with the future pattern of advertising locations in Germany, can be added when viewing the maps in a historical perspective: All Berlin offices have been established only since 1999. That is to say, as already suggested by the employment data, also from the standpoint of international network agencies there has been a very recent reorientation towards the new German capital. At present this does not seem to occur at the cost of Hamburg, but is more likely to threaten the present headquarter cities Düsseldorf and Frankfurt.

Also the present agendas concerning the new challenges differ paradigmatically between the two networks, regarding the re-orientation of production, labour strategies, location etc. McCann-Erickson's main strategic action taken within the last years was the establishment of the Berlin office promoted both as a historic 'circular statement' given that the agency had started there in 1928 and as a completely new agency conception, the 'McCann-Erickson Communication House (MECH)' which is to advance the concept of integrated communication under a special brand and in a particular city. In addition the re-orientation of training strategies did not focus on creative professionals but rather stressed the genuine assets of a global agency: internationality, integration, and marketing orientation:

> Yes we need generalists who know how communication works, a sort of superstructure which has an overview of integrated communication. [...] And as the market outside does not provide these people that we need, we are just starting a trainee scheme which we have tailored on our own.[...] Each of our eleven specialized companies employs one graduated person, and they have to rotate; they are based, say, four months here and then they have to learn what the public relations guys actually do and how the web-site people work; they stay three months in each specialized discipline that is, 18 months, and then they come back (personal interview, 2000).

Enhancing the creative reputation is thus not McCann-Erickson's main focus. Unlike that BBDO, although of course also highlighting integrated communication, strongly aims at increasing the local content of its business. The most recent action in this context was the recruitment of Springer & Jacoby's Chief Creative Officer for the same position in the BBDO group.[19] The objective was clearly, after having acquired minority shares in small creative agencies – above all in the Hamburg based KNSK, partly a Springer & Jacoby spin-off – to enhance the creative reputation also of the core brand BBDO which suffers from having been responsible for the old-fashioned advertising for detergents, in this context of course making use of the financial capacities of a global group.

Of course portraying the two basic strategic orientations as single alternative options would be a too straightforward argument. Achieving creative reputation, for instance, can also be determined by the specific group history as in the case of the WPP-owned Ogilvy & Mather that continuously varies between the ranks no.

---

19 *Springer & Jacoby* press release, 11/12/2002.

three and four in the creativity rankings of the 1990s, combining a centralized organizational approach with a 68 per cent share of national clients.[20] This is of course also due to the fact that entering creativity contests is an expensive endeavour which can be managed by big groups in an easier way. Also BBDO's de-central approach is by no means a pure orientation to the 'local content': The former S&J chief creative left the agency already after few months thereby reacting to a reorganization of the network on the European level.[21] Nevertheless the alternatives demonstrate that global advertising business in national markets involves an important dilemma and that there are paradigmatically distinct ways to deal with it which yet never will be able to provide a definite solution.

The Pioneers of Creativity: Different Strategies at the Threshold to International Network Agencies

Springer & Jacoby and Scholz & Friends today constitute two well-established top ten agencies on the national market. Both are groups with about 600 employees;[22] both are among the top three in the national creativity rankings, and both have begun to extend their activity outside the German territory, Springer & Jacoby presently with national offices in London, Paris, Barcelona, Milan and Vienna as well as an international headquarter in Amsterdam, and Scholz & Friends with a substantial network, including 16 European offices with a strong focus on Eastern Europe. Both agencies oriented their strategy of internationalization to their main clients: Springer & Jacoby with DaimlerChrysler's Mercedes-Benz account and Scholz & Friends as communicative support of the Eastern Europe extension of the Hamburg based Tchibo/Reemtsma-group. Thus, it was actually the classic motivation of 'following key clients' that drove them towards international markets.

In addition they thereby reacted to a fundamental growth limitation given that an increasingly global business in all sectors tends to require also the communicative support on a supranational level. In the case of DaimlerChrysler, for instance, this implied a shift from using the 'best local agency' in each national market to a more global approach: 'Only since 1990 Mercedes says: "We understand ourselves as a brand, even as a global brand and therefore we must make a real global brand communication"' (personal interview, 2000). The fact that the pioneer agencies were not able to provide a global service increasingly proved to be a problem, for S&J leading to the loss of multinational clients such as IBM already in 1995 (Boldt, 1996), but also of big national businesses such as the Lufthansa.[23] This challenge of increasingly global – or at least European – adverti

---

20 From http://www.ogilvy.de/ogilvy/index2.html, 19/1/2002.
21 *BBDO Germany* press release, 30/03/2004.
22 As a comparison: The whole BBDO Group had 2858 employees Germany-wide in 2000 (*Werben und Verkaufen*, 13/2001).
23 From *Horizont* online news, 21/02/2000.

sing markets also applies to the growing segments of the last years, that is, the deregulated public utility markets such as telecommunication, post and transport services, all having provided important clients for national agencies in the 1990s.

Yet, internationalization as a means to overcome these limits has brought about new challenges for the two agencies since they now run the risk to adopt the problems of the global advertising networks, that is, on the one hand, to lose the 'speed' and attractiveness of a relatively small enterprise by having to manage big accounts in different countries and, on the other, to need to culturally embed the own advertising in foreign markets. In addition, both challenges have to be faced while starting from a latecomer country in a global market actually occupied by big and experienced networks:

> But you have to learn the following: To everything there is a season, and if you miss it, that is, if you miss to build up internationality, you cannot just catch up, or you only can do it, and S&J only can do it by saying: 'We are a speciality!' (personal interview, 2000).

This dilemma was the reason that the founder generation of both agencies for a long time remained hesitant in expanding their activity. Whereas the next generation of the Scholz & Friends management has meanwhile notably fostered the abroad business, Springer & Jacoby's pace still has remained substantially lower, only from 2000 on successively establishing an office network in European core markets, strictly oriented along the key principles of the parent firm:

> And also our international network shall be built with the unit-principle, that is, in the different countries we shall take up local, young and ambitious people as partners, so that in Spain a Spanish leads the agency, as well as an Italian in Italy etc. so that we get a glowing combination of different nationalities, but oriented to creative campaigns, and if you succeed to develop creative campaigns you always have a chance (personal interview, 2000).

A second challenge the pioneers are confronted with is the increasing need for integrated advertising strategies including all potentially available communication channels. As both have emerged as classic 'above-the-line' advertising agencies and given that their main product innovations regarded chiefly TV advertising, they are not really perceived as serious actors in integrated communication although they are able to provide a comprehensive campaign in co-ordinated networks of independent specialized suppliers. This in turn presupposes clients tending (and able) to orchestrate accounts by themselves. The map in Figure 4.4 depicts again the DaimlerChrysler-case which interestingly includes both pioneers, Springer & Jacoby as lead agency and Scholz & Friends as service firm for commercial vehicles.[24]

---

24 In addition, the map provides some hints as to the changing locational structure of the Daimler Benz advertising in the last decades. There are still leftovers of 'traditional'

In addition both agencies undertake own efforts of establishing specialized suppliers of all communication tasks. S&J, as with internationalization, also in this context continues to emphasize its original and distinctive culture. In the words of one of its managing directors, related to the role of S&J's internet subsidiary 'Elephant Seven' which was set up as a specialized internet unit, after some time partly bought out by its management and in the course of the new economy-crisis completely leaving the agency group:

> It's logical that an agency brand like S&J deals with the issue of integration. Who does not do that? But we give much more importance to the question: 'In what way will this be able to work?' And there is actually no agency in Germany that has a really 'sparkling' answer. And, as regards E7: Participations do by no means give any guarantee for a powerful collaboration. It's much more important that firms and people stand for the same culture, the same values and tools. And in this sense S&J and Elephant Seven perfectly fit together (Kemper, 2002).

Also in this area Scholz & Friends revealed a less cautious attitude, already in the second half of the 1990s setting up specialists in sponsoring and event marketing, public relations, market research and design, in the meantime 'branding' themselves under the heading 'The Orchestra of Ideas' (Scholz & Friends, 2004). In addition the agency merged with a television entertainment firm in order to exploit new synergies between advertising and TV programming thereby considering themselves to be 'at the forefront of the convergence' of different media sectors, by developing new and 'complementary ways of communication that involve the audience such as TV-events etc.' (Scholz & Friends, 2001). Given that the new partner had been quoted at the stock exchange the merger also opened the stock market for S&F as a new capital source in order to support the widening of the agency network.

That is to say, the pioneer agencies differ plainly as regards both depth and pace of their reaction in the face of the challenges of globalization and integration. Interestingly, each of them shows an apparent paradox in its trajectory: On the one hand, S&J, which most strongly focused on innovation and change when relating itself to the traditional German advertising, now proves to be rather conservative and cautious. The agency appears to be afraid of losing the base of its innovativeness when advancing too fast. In the same way as the growth of the agency has been organized very carefully, constantly keeping in mind the basic philosophy of Springer & Jacoby, both internationalization and integration are to be oriented along the distinct agency culture.

---

suppliers in the region around Stuttgart where both the headquarters and the main plants of Mercedes-Benz are located. This structure has been 'disembedded' from a downstream-orientation and 're-embedded' in an upstream perspective, obviously of urban environments attractive for the relevant labour force. Also the marketing department has followed this logic.

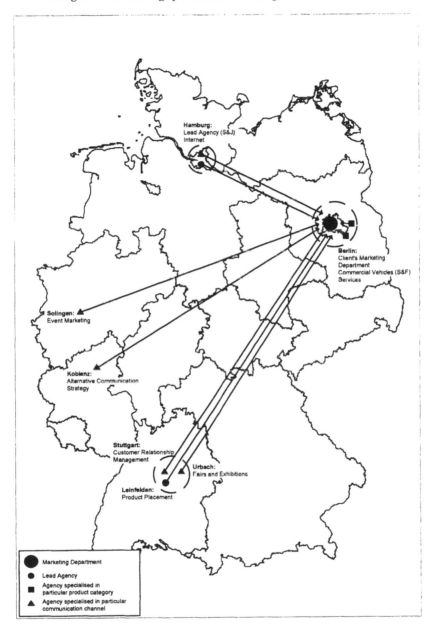

Figure 4.4   The Locational Structure of the DaimlerChrysler 'Star Network'

Sources: Kemper et al., 2000; cartography Frauke Funk, Andreas Beekmann

On the other hand, S&F, having actually been a reorganized product of the classic brand advertising of the 1970s, appears to be much less restrictive in widening its activity to new fields. This applies for the geographical extension of the business, the integration of communication channels, new sources of finance and the agency's locational strategy, the latter visible in the strong emphasis on Berlin to where Scholz & Friends meanwhile has moved its headquarter. For Springer & Jacoby, despite meanwhile steering international campaigns from Amsterdam, Hamburg, and even the address 'Poststraße 14-16' as the agency's 'germ cell', remains a vital element of the own firm culture, the specific narrative of an agency explicitly different from the traditional models:

> There is also only one Eiffel tower. And there is only one Springer & Jacoby; it is unique, and it is in Hamburg, and given that Germany is so small and that you can travel to any place without any problem, that is enough, isn't it? (personal interview, 2000).

Whose strategy proves to be successful is unclear. Both agencies have had some problems in the course of the new economy crisis since mid 2001. S&F in the meantime re-bought nearly all of its stocks so as not to remain exposed to a turbulent stock market – at the same time however buying also the portion of the British Cordiant Communications Group, the main shareholder which owned the majority of the agency already since the middle of the 1980s (Scholz & Friends, 2004). Springer & Jacoby during the last years became above all victim of its own success in that it lost a lot of important management staff, to BBDO, as in the case of the creative head André Kemper but also to other (not only) international agencies or even to client firms. For instance, Michael Trautmann, the longtime account manager for the key client DaimlerChrysler, in 2002 became head of the Audi marketing.[25]

Interestingly, however, both cases of 'people drain' from S&J cannot be understood as heralding the end of the success story triggered by the pioneer agencies and their mobilizing internal labour market strategies. On the contrary, the Springer & Jacoby-people failed in the organizational context of bigger and more traditional firms. They finally decided to set up their own business, a new agency named 'kempertrautmann' which directly after its foundation succeeded in acquiring one of the big national consumer electronics retail chain as founder client.[26] Thus, in the end, the example paradigmatically stands for the reflections the pioneers' internal labour strategies have on the external labour market, changing the framework of competition and, ultimately, leading to the emergence of a growing group of small and fast growing agencies, the second generation of creative advertising in Germany.

---

25 *Werben und Verkaufen* online news, 05/04/2002.
26 *Werben und Verkaufen* online news, 01/09/2004.

The Second Generation: The Founders' Reputation and the Ambivalence of Acknowledgement and Distinction

Out of the biggest 200 advertising agencies on the German market in 2000, 125 were set up later than 1981, that is, after the establishment of both pioneers.[27] That is to say, a large portion of significant players, although not the really big ones, in national advertising has been established during the last 20 to 25 years. Of course not all of these start-ups can be considered as direct consequences of the dynamics triggered off by the pioneers. In some cases, for instance, the new agencies were established as affiliates of international networks. Other cases do not appear in the ranking any more since they have either gone bankrupt or merged with a global group. However, it is clear that in German advertising, as happened throughout the world, the period since the early 1980s has been strongly characterized by the emergence of new players, many of them stimulated through the changing image advertising had through the 'creative turn' inherent to the 'second wave'.

In this context, one can argue with a certain legitimacy that the basic framework shaping the activity of the start-ups following the pioneer agencies more or less equals the situation the originals were facing around 1980: A group of founders decides to go out of their present jobs to set up an own business. The new agency has to be positioned as a player in the marketplace; reputation has to be built both in the client sphere and on the labour market; the internal processes of the firm, including the management of growth, have to be assured. Unlike Springer & Jacoby and Scholz & Friends, however, these start-ups are established while the 'pioneers of change' already exist. Thus they operate, on the one hand, in an atmosphere open to a new and creative advertising since the originals have so to speak prepared the ground for it. On the other hand, they even more have to justify their very existence given that it is not clear that they are better than the originals.

How this process of building up an advertising business works, aiming at being acknowledged as serious but simultaneously unique players in the sector, or, at being and not being like S&J and S&F, depends strongly on the personality of the founder(s), respectively on their professional biographies and often on the way how these biographies are interwoven with the pioneers. The most impressing and most successful example is the agency Jung von Matt (JvM), set up in 1991 by two managing directors of Springer & Jacoby, taking both their founder client and 15 people labour force with them and becoming the second most creative agency in Germany during the 1990s. In the meantime they constitute more a 'second S&J' than a member of the second generation.[28] Figure 4.5 depicts the careers of the agency founders, and above all how they are 'filtered' through the German advertising system, in terms of both agencies and cities.

---

27 Counted from *Werben und Verkaufen,* 13/2001.

28 JvM also follows the pioneer regarding its client structure, working for the second German luxury cars group BMW as well as two of the big former public monopolists, the post and the railways (Deutsche Post AG, Deutsche Bahn AG).

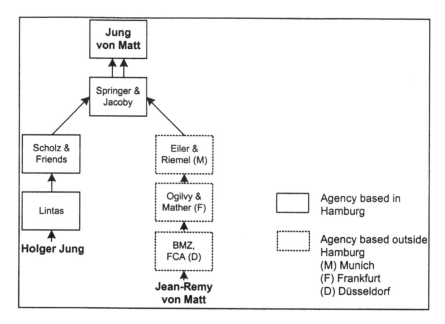

Figure 4.5   The Professional Biographies of the Jung von Matt-founders

Sources: Jung, 1999; von Matt, 2001, 2002

This personal professional history, on the one hand, drives the decision to become independent, through either positively motivating employees or negatively making them unsatisfied with their job in the environment of a big enterprise:

> We were fed up with doing the 25th same international job for Unilever, Nestle or something else. We did no longer want to sit in endless meetings with 20 people from 15 countries discussing a strategy or something, finally ending with the statement that it was good that we met, thus without any result (personal interview, 2000).

On the other hand, the founders' professional careers also affect the performance of their 'new' agency by both shaping their personal networks and providing them with the necessary reputation they need to successfully perform.

Thus, the former work experience, first, helps to generate a reservoir of contacts to be used to acquire clients, recruit labour etc. In this context start-ups are often based on existing work constellations among members of the agency management and between them and the management of client firms. The way Jürgen Scholz built his agency on his friendship with the owner of the Tchibo-Reemtsma group can manifoldly be found in the establishment of the agency generations following to Scholz & Friends. Sometimes this mode of foundation is triggered by conflicts caused through changes in the client structure of the parent agency. But also the common work experience within an agency is helpful for creating a management team for a start-up:

> You have collected profound experiences you can use. And – most importantly – you have found a constellation that works. There are lots of start-ups on the marketplace, where people just meet in a pub and say: 'Let's do it together'. Therefore so many agencies are renamed every year: 'Schmidt and Partner', 'Schmidt and Müller', then Schmidt leaves, it's 'Müller and Somebody' etc. (personal interview, 2000).

Second, the agencies which founders have gone through before starting their own business as well as the positions they have held within these agencies significantly contribute to the 'standing' of the new firm. A Springer & Jacoby management experience, for instance, of course represents a specific approach to advertising, but also a certain ambitiousness as regards the quality of the own work, thereby facilitating the own 'branding', indispensable for a successful performance.

> You need to have a name. An agency without a special name is nothing. There is a whole body of agencies that certainly do a good job. But nobody knows them, and they therefore do not get big clients, nor are they able to attract good labour force (personal interview, 2000).

Generally put there are two start-up patterns which were already reflected in the establishment of the pioneers: an 'opportunity'-variant based on existing client relations, frequently triggered by client conflicts of the founder's former employer, and a 'challenge' variant constituting a more radical break in the personal career of a founder. The following quotation depicts an extreme case of the latter:

> It is always a risk to set up a firm [...]. And we did it without taking a client with us, since we explicitly wanted a new challenge. [...] We were in the appropriate age, not being a youngster any more but also not being exhausted [...]. We dropped all things that had made us lazy: a comfortable car, a comfortable income, a comfortable staff arrangement, there was nothing of it. Too much administration makes us slow; too many big cars make us fat. We really try to live that idea. It's a sort of outlook upon life (personal interview, 2000).

Of course every firm foundation involves both a challenge and an opportunity. However, given that the 'opportunity'-model generally already starts on a higher level, the very beginning of an agency's existence is likely to influence the later trajectory of the firm. This is not to say, the spin-offs based on a founder client are more successful. Three of the most thriving German start-ups in the 1990s began more or less as underdogs (Kolle, 2001). Regarding the conflict between profiling and acknowledgement one can even argue that the 'opportunity'-model facilitates rapid acknowledgement, yet at the same time complicating to develop an own profile.

In this context it is not the key problem to run an agency profitably at all but rather to surpass the threshold to becoming acknowledged as serious players in the national market, either by having 'blue chip' clients or by covering the whole range of relevant market segments:

> It is generally considered that there are, in terms of sector diversification, three decisive disciplines in advertising: cars, tobacco and beer. Cars are technology. Cigarettes constitute a purely virtual advertising world, since nothing is done through the price, there is actually no innovation; it is only image that matters. All other marketing parameters can largely be left out of consideration. And the third is beer, as you have to deal with the food retail system, catering trade, event organization and the specialized beverage retail trade, very different and specific communication channels that confront the agency with completely different requirements (personal interview, 2000).

In our judgement there are basically two key qualities agencies must have in order to increase their reputation: First, it is again the reputation and the contacts of the founders which determine the agency's performance. Second, reputation can also be built in the field of creative success. Particularly international awards are – even if still of minor importance on the client side than other parameters – increasingly a 'currency' in which advertising agencies are measured and through which they are able to promote themselves. The example of one of the most successful agencies of the late 1990s in Hamburg shows how a *Cannes Lion* could push its performance:

> Well, one milestone for us was the proof that we are a creative agency through an award in Cannes, on this advertising festival. That award immediately triggered a wave of PR for us. That was incredible. These no-names suddenly turned into such, such well-known guys, and at the same time we then got invitations from bigger clients like XXX and brands you would normally not place in such a small agency (personal interview, 2000).

This implies however the major importance of having reliable high quality labour force just at the beginning of an agency life, in turn implying the particular need for a good reputation and good contacts in the labour market:

> Of course, the relations I had were on the client side. And my partner's relations were to other agencies, because the recruitment of personnel [...] is clearly an important issue, particularly for an agency of the size, which we had at that time. For we wanted to think and work out 'first-league ware' on the highest level; but for 'first-league ware' you need 'first-league players'. But a 'first-league player' only joins a 'first-league team' what we were not, at least not on paper. So people had to be convinced [...] and since my partner had a good reputation in the scene [...] we were able to commit people to us and to get the freelancers, which can of course chose where they want to work, to assist us in our campaigns, in the development of projects which we would not have managed alone (personal interview, 2000).

That is to say, the second generation agencies from the very beginning of their existence need an active and similarly professional labour strategy as the pioneers including the continuous screening of the labour market, the development of training schemes. Even more, also and particularly on the labour market the start-ups have to justify their existence given the attractiveness of Springer & Jacoby and Scholz & Friends. One important argument in this context is smallness alone:

> It's of course much more interesting to contribute to the moulding of an agency like ours. That is the nice thing here: We still have a size at which every employee, even the cleaner, is shaping the agency since the influence you have is immense, of course particularly for management staff. And then I can consider: 'Do I want to be the, say, 35th executive of S&J, there certainly doing a good job, or do I really want to leave an own footprint here?' And this is here rather possible than there, even if you earn less (personal interview, 2000).

However, profiling on the labour market is a much more comprehensive issue, including the work atmosphere and environment and, importantly, the location of the agency within the city. Thus, whereas global players and the pioneers tend to locate in the central parts of the city, preferring representative addresses or reflecting the continuity of a specific firm history, the second generation players, beyond aiming to have enough and cheap enough office space to grow, very much intend to 'communicate' their specificity and uniqueness through the location decision. One example of an agency in an old protected warehouse ensemble in the Hamburg port area:

> This is the absolute non-plus-ultra. We were among the first here at the waterfront, in the old part, not where the yuppies are. And we find it superb. Our people feel fantastic, [...] I feel fantastic. It's simply nice. That's unique in Germany and this uniqueness is important. It's also not a sort of museum but it is the original – simply superb (personal interview, 2000).

The Traditional National Players: The Threat of the Post-War Model vs. New Specialist Opportunities

Seen from a present standpoint the category of 'Traditional National Players' appears to be the most hybrid one given that it is ultimately determined by historical criteria. This is to say, it is difficult to identify typical examples of a nationally dominated traditional agency since this group has been the one that was most strongly forced to change throughout the history of German post-war advertising. Based on a very rough attempt to quantify this group, among the top 100 agencies in Germany there are 29 owned by a national majority and having been founded not later than 1980. The first of them holds position no. 15. When extending the ranking to the top 200, the figure rises to 61,[29] however being likely

---

29 Counted from *Werben und Verkaufen,* 13/2001.

to include agencies with a strong focus on regional clients. In addition, traditional national agencies have frequently functioned as gateways for global networks to enter the German market, either through complete mergers or through the acquisition of minority shares. The assumption that this segment is very likely to disappear in the near future therefore seems realistic, particularly given the dynamic development of new and 'sparkling' protagonists of the 'creative turn'.

Yet reality seems to be more complex and the competitive environment of the traditional German agencies can only be unravelled when taking their specific history into account. They were generally established in the tremendous growth period after World War II, their growth thus being based on the national mass consumption clients of that time: food, fashion and other fast living consumer goods. In addition, their room for manoeuvre was restricted through the fact that the American advertising 'factories' constituted the big players on the market. These basic conditions brought about a series of characteristics for the nationally owned part of German advertising: a specific client structure, usually based on traditional small and medium enterprises, a specific post-war generation of entrepreneurs both as agency founders and as clients and, finally, a specific type of advertising production, focusing on printed brochures and catalogues and less on mass media. This particular starting environment is still mirrored in the agencies' business today, including all relevant issues, even the present agency location which frequently depends on the founder's previous choices regarding real estate investment.

There are three evolutions through which this model has come under pressure in the course of the last 25 years: First, the client base has been reduced due to concentration processes in the corresponding industries. Small regional food brands have disappeared, respectively have been bought out by global players like Nestlé, Unilever etc., these in turn tending to reduce the number of brands produced world-wide (personal interview, 2000). Second, the main growth areas have changed, regarding both the client segments and the media mix. Consequently traditional print-oriented agencies lose ground in the marketplace but also regarding their image in the professional milieu with consequences chiefly on the labour market:

> And we have never rated high in the Hamburg advertising scene; we were always an agency about which people said: 'A specialist. [...] Let them make their trade folders and brochures.' But we did always earn money (personal interview, 2000).

Third, the founder generation has begun to leave the business and it is an open question whether the transition to a new owner generation is successfully managed.

On the other hand, due to the rediscovery of non-classic tools within strategies of integrated communication the most recent years have shown a counter-trend particularly to the second aspect. Specialized agencies have become the fastest growing ones on the German market.[30] With this wind behind them, also the traditional national players have been able to exploit their formerly depreciated capabilities even being able to acquire 'below-the-line' accounts from new sectors such as household technology, financial services etc. In addition also the classic client segments have remembered the importance of a communication along the distribution channels. It is thus the specific competences their traditional clients had required that have now turned from an obstacle into a new growth potential in a glutted media system. Partly even international advertising networks have discovered this potential acquiring minority shares in traditional agencies with specific foci both on the German consumer market and on how to use 'below-the-line' communication advertising on it.

That is to say, the future of the traditional owner-led advertising agencies by no means consists definitely in their successive disappearance; however, given that also global agencies and the pioneers increasingly establish or buy communication channel specialists, the future growth will not come about automatically. One key question will be how the generation change in ownership and management will be undertaken and whether it is managed to combine the traditional assets in terms of client base and productive capabilities with a new organizational structure attractive for ambitious labour force. That is to say, also for the traditional national players, the labour issue will constitute one of their key challenges for their future performance. Whether their broad experience in training juniors within the framework of dual system professional education in Germany is useful in this context remains an open question. Of course this experience can be exploited; yet this again depends on the organizational structures it is embedded in.[31]

The Regional Players: Market Growth and the Limitations of a Regional Client Base

When seen in terms of firm numbers, the major part of the advertising industry can be assigned to the regional players, that is to those agencies focusing on regional clients and regionally oriented communication strategies. Just to give a hint as to the quantitative dimension: The firm database of the Hamburg Chamber of Commerce in July 2001 counted 3,395 classic advertising firms.[32] Among the top 200 on the German market there were however only 30 Hamburg based agencies.

---

30 The year 2000, however, constituted a notable exception since the new media euphoria fostered an extraordinary growth of classic advertising (ZAW, 2001). At the moment, expenditure for 'above' and 'below-the-line' advertising is equal. Experts however forecast that 'below'-spending will amount to two thirds of total advertising expenses already in the near future (Diekhof, 2001).

31 For instance, dual system training of advertising professionals and media production specialists is an inherent part of Springer & Jacoby's labour strategy. For many of the big advertising agencies it serves as a cheap means to recruit young and talented staff (personal interview, 2003).

32 From http://www.hamburg.ihk.de/produktmarken/produktmarken.htm, 10/09/2001.

Even if one rightly questions the reliability of both data sources it seems legitimate to hold that advertising players linked to their regional environment in a downstream-logic constitute a big and important fundament for the overall sector, mainly as regards their role as training sites, flexibility reserves etc. on the regional labour market. Yet, the market conditions in which they operate differ essentially from the rest of the industry. Four aspects are worth being mentioned:

First, the 'creative turn' has affected the business on regional markets only to a limited extent, if at all. Indeed the overall requirements on the artistic quality of advertisements, catalogues, brochures etc. have substantially increased in the course of the last decades; yet this has rather contributed to an increasing professionalism of the sector as a whole and to the growth of free-standing advertising firms to the detriment of in-house market communication than to a new advertising style.

Second, this in turn implies a general growth of the regional market.[33] The way how firms present themselves to the external world has turned to be a major tool for competitiveness, even for small and medium sized enterprises. More and more firms, but also public authorities as well as non-governmental organizations tend to a more professional market communication thereby increasingly using advertising agencies.

Third, this seeming sustainable growth is at least partly undermined by the fundamental technological changes the production process of advertising has also undergone in the last decades. The meanwhile total computerization of design has restructured the division of labour within the advertising value chain, permitting to merge finished art work, formerly carried out by stand-alone firms, and creative development. In addition digitalization has lowered the barriers of entry into the advertising sector given that relatively low fixed costs of hard- and software and the possibility to cover a big portion of the value chain with this cheap equipment facilitates to set up own businesses, even at home. At the same time also the other players in the value chain that is, printing shops and lithography firms, are enabled to diversify into computer graphics and desktop publishing (DTP) in order to provide their regular customers with additional services (Gonzalez, 2001).

Fourth, their clients – although being highly diverse – feature one common element in that they all have a low affinity to advertising as their core activities do not include professional market communication. In addition, their firm organization and enterprise culture – even more than in the case of the traditional food and fashion industry – are generally focused on the personality of the firm owner. Advertising is thus less the product of an encompassing marketing conception but expresses the perspective and the preferences of one single person. At the same time it is likely to include much more than classic (print) communication tasks such as the support in daily communication through business letters etc. Correspondingly also client acquisition by agencies or agency hiring by a client occurs in a 'hands-on' logic on the part of both partners:

---

33 See Chapter 3, p.67.

We get our clients in a very classic way. We have a comparably conspicuous advertisement inserted in the 'Yellow Pages' [...] and we have got remarkable clients through this. Hence, when somebody needs a poster or something like this, and he does not have anybody in mind who can do this, he takes the 'Yellow Pages' and the first advertisement he looks at is ours. And if we are located close to him, he may call us (personal interview, 2000).

In sum, regional players' strategic action is shaped by these clients thereby being largely hindered to develop a reputation necessary for advancing to the wider national market of mass media advertising. The result is, on the client side, a relatively rigid segmentation of the regional market. There are of course activities of young second generation agencies and traditional national players with regional clients; these are however usually of high public interest, e.g. public utility companies, local mass media etc. On the labour force side, as well as on the office market, the small regional agencies compete with the other players (as to the latter even with those from other service sectors). That is to say, the nature of the advertising client structure on regional markets does not only cement small agencies' position there; it simultaneously weakens it particularly in the labour market, given their relatively low financial power and the comparably little attractiveness of work in terms of creativity standards:

In the realm of creative development labour fluctuation is particularly high. This can be for two reasons. As regards not very talented people the creative capacity is used-up after two years. Talented people however tend to use our agency as a springboard into the labour market of big agencies. Given our client structure, our limitation to print advertising and the lack of individual promotion prospects we do not have any means to keep them. So with our last art director: We were really keen to keep him here but it was impossible (personal interview, 2000).

Advertising between Business Service and Popular Art: A Second Interpretation

In the first part of Chapter 3 we labelled the basic activity of advertising agencies as 'selling ideas', emphasizing that it involves two different facets: creating ideas and selling them to business clients.[34] We held that this double-faced nature of the industry is reflected in the functioning of the agencies' core activities account management and creative development. This core was considered to function like a 'hinge' that both orchestrates different tasks in the advertising process and bridges the different 'worlds of action' inherent to business and creation.

The preceding analysis of continuity and change in the German landscape of advertising agencies has confronted this general view of advertising agencies with a differentiation of agency types, each of which operates in a specific framework of competition in the national marketplace, and in the light of new challenges posed

---

34 See Figure 3.3, p.50.

by the new players of the 'second wave'. The following paragraphs are to re-synthesize this segmented view in order to obtain a comprehensive view of the mechanisms lying at the heart of both the differentiation and the interdependence in the overall industry, as well as to figure out the basic mechanisms behind the processes of change it has undergone in the course of the last about 25 years.

Basically this synthesis encompasses three dimensions closely linked to the just mentioned 'hinge function' of the advertising agency: first, its role to orchestrate different interfaces and the patterns of relations and institutions these are based on, second, the need to bridge the seemingly conflicting action frameworks of business and creative development and the impact of a changing relation between them, and, third, the role of place, visible in the distinct but empirically converging function of office location within the agency strategies.

Managing Complexity: The Nature of Relations and Institutions in the Making of Advertising

The function to orchestrate different tasks within the process of developing advertising campaigns basically means that an advertising agency has to deal with a multiplicity of interfaces within and, above all, outside its borders, from the client as the one who pays and thus ultimately determines the focus of the strategy to the director of photography of a commercial or to a freelancing art director involved in the development of a particular campaign. In our view the main structuring variable in this context is that the campaigns and, consequently, the co-ordination tasks within these campaigns differ in terms of their complexity. The following Figure 4.6 is to illustrate the logic of high and low degrees of complexity drawing on the typology of agencies developed as starting point of the interview study.

Roughly said the complexity of campaigns thus tends to increase along the diagonal line of the figure, that is, it is determined both by the size of the market and by the 'style' or the creative or 'artistic' standards of the advertising. Increasing complexity means that both the number of agents involved, and thus the number of interfaces to be dealt with, and the quality level of tasks rises, be it in terms of its cognitive or artistic demands. It depends on the media used, on the size and complexity of the market, on the market position of the advertised product etc.

In the light of the evidence described above it is not only the number of agents and the complexity of their contributions that changes with the changing complexity of advertising campaigns. Also, and more importantly, the nature of relations that can be found between the agency and the further agents involved also strongly varies according to the complexity of a campaign. Again said in very general terms: The higher the complexity of the advertising process the more strongly the relations between agents are socially underpinned, respectively the more strongly they have the character of relations within a professional milieu involving common conventions, common standards for the building of reputation, commonly acknowledged protagonists within the industry, a high density and high likelihood of personal relations etc. This affects three kinds of interfaces: to clients, to suppliers and co-operation partners in the production process, and to labour force.

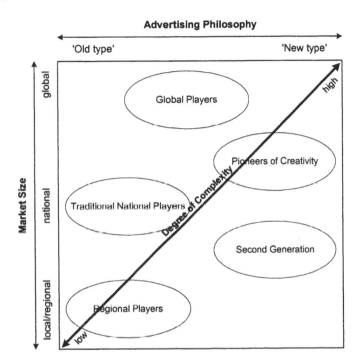

Figure 4.6   Increasing Complexity of Advertising Campaigns by Type of Advertising Agency

Sources: Own illustration

To give an illustration: The relational environment of the regional player, whose activity displays the lowest degree of complexity, is involved in is generally characterized by a relative anonymity and casualty. Clients are acquired either through neutral market communication in the corresponding media, such as the 'Yellow Pages', or via personal relations within the client sphere, that is through recommendation from one client to another. In the best case a relation is built through an active 'new business' strategy by which potential clients are to be convinced through quality, as in the case of a self-employed graphics professional:

> We nearly do not advertise for ourselves. It's all verbal propaganda. What I sometimes do, however, is to make alternative layout proposals to existing ones. I did it for instance for an old people's journal, first making – for nothing – a new layout since I found the old one terrible. Then I presented it to the boss of the journal. He loved it and gave an order to me (personal interview, 1997).

These once established relations generally tend to consolidate, in practice however still being based on casualty given the fact that advertising is not part of the clients' everyday business. A similar logic characterizes the relations both to the labour market and to suppliers and partners for co-operation, except the long term and usually stable, but little complex linkage with printing shops.

At the other extreme of the diagonal line, that is, between the pioneers and the global players, relations at the interfaces are generally based on a high degree of common knowledge as regards the different players and their situation, or at least of resources that enable actors to quickly obtain this knowledge. Client acquisition – unless happening on the global level and affecting a national agency office without any own involvement – in a first step occurs either through personal networks between management staff of agencies and clients or through reputation based on institutions in the sense of 'congealed social networks' (Martin, 2000) that is, on common conventions detached from the direct personal relation and sustained through an institutional environment that consists of specialist trade journals, sector events and meeting places which transfer sector knowledge and create and circulate 'stories' and 'narratives' (Amin and Thrift, 1992). This general reputation involves, respectively is associated to a certain distinguishing agency image which obviously strongly varies with the agency type:

> Well, imagine a client who intends to place an account and screens agencies. One key element in this context is the GWA-annual[35] that provides a really good overview of different profiles of agencies. And if you are in the mindset and if you have a clear profile within this mindset you really have a chance to get a call and to be invited for a pitch. As regards profiles there are some agencies that are considered to be particularly creative […], others rather have a reputation in strategic marketing […], and there are some that are somewhere in between. And then an executive says: 'Let's take one of each group and test them. We will see which will be the most appropriate for our needs' (personal interview, 2000).

In a second step clients test whether this combination between the general reputation and specific profile of agencies proves to be applicable to their communication needs, either by competition in a pitch or by a limited project account that is to test an agency under conditions of real co-operation.[36] Thus, given that awarding an account contract to an agency does not signify to buy a clearly defined, 'tangible' and testable product, clients aim at reducing uncertainty as much as possible both by constantly being involved in the agency landscape and directly testing single agency's capacities to work for them.

---

35 Gesamtverband Werbeagenturen (German Advertising Association).

36 The importance of project accounts is increasing, particularly for young and small agencies. However, while they create an opportunity to earn reputation through clients bigger than actually expected for a small agency they are more demanding for the agency's organization as they destabilize the income of an agency (Gieseking, 2000).

The logic of reputation plus profile also works in the inverse direction. Also clients differ in the way they are seen by advertising agencies, on the one hand concerning their general importance in the market that is, whether or not they are 'blue chips', on the other as regards their affinity or openness to innovative styles of advertising. This 'profile' dimension is frequently also reflected in the way how clients tend to work with the agency, whether they attempt to determine all decisions within the campaign or whether they trust in the agency's capacities to accomplish their market communication in an appropriate way.[37]

That is to say, agency-client relations within complex advertising processes can involve positive feedbacks driven by the institutional and conventional environment of the sector within which reputations and profiles are attributed. The interaction between agencies' and clients' profiling strategies can stabilize market segments based on different advertising styles. The key issue in terms of a substantial innovation is the question of 'blue chip' clients. When Springer & Jacoby succeeded in winning the Daimler-Benz account they managed to 'generalize' an entertaining advertising style. We shall come back to this in the next subsection.

On the input side, that is, as regards relations to the labour market as well as with potential partners and suppliers within the process of production the nature of relations is on the one hand similar unless partners or suppliers are part of the same wider group as the agency itself. As to suppliers of standard technical input such as lithography, for instance, agencies pursue a strategy comparable with the selection process clients use when screening the agency market, combining elements of reputation (or personal experience) with elements of competition:

> Our producer has contact to four lithography firms [...], of course we have one we prefer maximally with which we co-operate more than with the others. But there are three further firms, to maintain competition. Thus we don't want to depend on one single supplier. And when one of them does not work, for whatever reasons, he will get another chance, but then he will be switched off for a certain time. That already happened. You can certainly imagine how hard they tried to fulfil the next order perfectly (personal interview, 2000).

In the case of more demanding tasks such as directing a film the suppliers' reputation is usually more significant for the final selection than personal experience.

Generally the interfaces on the input side are more complex both due to their mere quantity and to the fragmentation of the field of potential suppliers for each specific need. This in turn implies that the role of institutionalized reputation is less important than direct personal experience, respectively personal relations

---

37 Schmidt and Spieß (1994, p.91ff.), drawing on an the evidence from analysing S&J's production of a TV-commercial for a big mail-order group, argue that one of the key presuppositions for a quality of the film satisfying for all involved actors was that the relation between client and agency was characterized by the client's trust in the agency's competence.

through which trustworthy information is provided. The labour market is a particular example in this context, due to its logically extreme fragmentation, but also due to the very nature of labour as a 'fictitious commodity' and the high degree of uncertainty linked to this nature. The problem of not buying a testable and tangible product valid for a client's decision about hiring an agency even more applies to the recruitment of labour. There are of course both acknowledged quality and qualification standards, which can be used for the selection of staff, and mechanisms of reputation building such as creativity awards or informal information about who was responsible for certain successful campaigns. However they do not guarantee the successful performance of a given, say, art director in a particular work environment.

In sum, the nature of relations and institutions changes with the complexity of advertising, in very rough terms, from casualty and formality to regularity and informality. This difference is mainly owing to the stronger need for trust in complex environments as well as to the fact that the mechanisms of trust generation vary with the degree of complexity. However, just looking at this static and binary logic of high and low complexity tends to ignore important mechanisms within the structure of the advertising landscape. Above all, it does not provide any clue for understanding the restructuring of the industry, except suggesting that the importance of social and institutional aspects has increased given the rise in complexity.

Thus, it is necessary to look beyond the mere nature of relations and institutions stressing the differences between the client sphere and the input sphere of an agency not only owing to the greater fragmentation of the markets. The fundamental cultural barrier between 'client-liaison function' and creative development in an agency does apparently determine the institutional-relational pattern it is embedded in. The following quotation shows an example of how a certain market sector related reputation in the labour market cannot surpass the barrier to the client sphere:

> We do not have any client in the field of technology. Clients, when looking for orientation in the agency market, are using references such as 'they have worked already for Audi' or so. [...] And it is nearly impossible to convince them that you can do it. [...] We have people here that have been responsible for Mercedes-campaigns, Audi-campaigns, BMW-campaigns and Lada-campaigns; the people who conceived all this have meanwhile come to us. But in the market the agency brand does not have any competence in the field of technology (personal interview, 2000).

That is to say, it is in terms of the double-faceted nature of advertising that most of the industry's structuring and restructuring has to be conceptualized. We shall in the following section try to unravel the logic of this basic and unique separation of an industry and to figure out how it contributed to the process of restructuring of the last 25 years.

Bridging Different Worlds: A Model of Advertising between Business Service and Popular Art

Lash and Urry, in their account of the culture industries as pioneers of a both reflexive and flexible economy, argue that contemporary advertising so to say constitutes the paradigm for the increasing convergence of the spheres of culture and economy (1994, p.138). Taking Saatchi & Saatchi's self-description 'commercial communications' as starting point they hold that this label well displays the two dimensions of the business – 'commercial' connoting 'industry' and 'communications' connoting 'culture': 'Advertising in effect evolves from a free-professional type business service to, in Fordism, an industry and, in post-Fordism, to a fully fledged "culture" "industry"' (ibid., p.139).

This characterization of advertising as a two-faceted business on the one hand appears to be compatible with our understanding of the 'bridging' of different worlds. However it gives the impression that this double identity works without frictions. Unlike that we consider precisely the 'dissonances' between the philosophies of advertising as business service, aiming to solve their clients' problems of market communication, on the one hand, and of advertising as popular art, on the other, to constitute its crucial feature that substantially shapes firm strategies, labour market mechanisms as well as the territorial structure of the business.

Figure 4.7 represents an effort to model this dual logic of advertising. It can be read along two axes: The diagonal line displays the conflict between the two 'worlds' outlined above. The advertising agency is in the middle of both trying to 'harmonize' the different philosophies. Its performance as a business service chiefly depends on its business reputation acquired – roughly said, and in this context abstracting from the logic of profiling and differentiation outlined above – through successful work for well-known clients. This system of growth is characterized by a self-reinforcing mechanism which is depicted by the bold circulating arrows. The relation is simply 'the more clients you have, the more you get.' In this sense, advertising does not differ from any other producer service such as consulting or engineering.

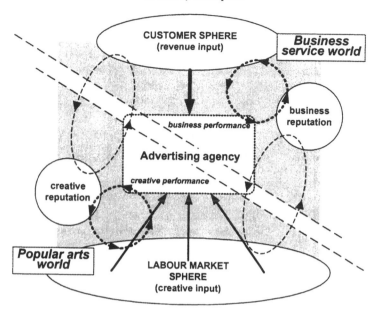

Figure 4.7   **Advertising Between Business Service and Popular Art**

Sources: Own illustration

The unique features of advertising lie in the other world, in its character as a popular art: 'Creativity' as the catchword within this universe does not just serve as an input to the solution of problems in market communication but it is so to speak 'institutionalized', in the form of national and international creativity contests similar to other culture industries as movie production and theatre. Winning a *Clio,* a *Cannes Lion* or similar awards may belong to the biggest successes an agency can achieve. High positions in single important contests and accumulated success in creativity rankings improve the creative reputation. Also here, a positive feedback of self-reinforcement is at work, however mediated through the labour market. As the motivation of creative and artistic labour is driven chiefly by their acknowledgement in the artistic community, high creative reputation attracts creative labour force which again is likely to improve the creative performance of the agency and so on. This circle also works in the inverse direction, since creative professionals responsible for awarded campaigns substantially increase their reputation.

Put in another way, the diagonal axis of the figure shows the different basic motivations as well as 'performance criteria' (Girard and Stark, 2002) driving the business service world and the popular arts world of advertising.[38] These different

---

38 Grabher, in his work on the British advertising industry, widens this model by introducing a third sphere, 'the scientific logic of advertising', represented by account planning (2002, p.248). See Chapter 3, p.51f.

motivations are reflected in equally different ways in the two main interfaces of the agency, the customer and the labour market interface, respectively. In this dual logic the bridging of these two worlds by the agency signifies to channel creative labour into business success and to convince customers that creative reputation is a necessary tool to attract talented labour force and to motivate it for good work. Or, in the words of an agency boss:

> ... all those contests, as businessman I would abolish them, as strategist of course not because the creativity contests are nothing more than [...] wind behind the creatives [...]; that's competition, and if there were no competition the party would be poor. [...] So you need some cocktails, and the cocktails are the awards [...]. I always said to the customers who are always sad because of not winning anything [...]: 'You are silly, the award is to get good staff, and only if you have good staff, you client will be able to get good work. And therefore you must be keen on your agency getting awards, not for you, you can throw them away, for your turnover they are also just of secondary interest, but you need good staff, that's logical, isn't it?' (personal interview, 2000).

However, bridging the two worlds is neither easy nor even self-evident. There has been a long and intense controversy in the advertising world about the criteria to measure creativity and, thus, to judge the quality of an advertising campaign. 'Clearly, winning an award for creativity does not guarantee success for an agency' (Wells et al., 1989, p.325). Although the creativity awards influence the industry in various ways,[39] creative reputation only enhances business success if the positive impact on the promoted product can really be proved. As mentioned Springer & Jacoby was only accepted as a 'serious' advertising agency when winning the Daimler-Benz account in 1989 and achieving measurable success in sales increase. This was five years after having won their first golden *Lion* at Cannes and despite having been the most creative German agency a long time before (personal interview, 2000). Thus, the transfer of creative reputation into the business service world presupposed not just to convince a 'blue chip' client but above all to prove its value in terms of business standards.

The change which has taken place in the industry has been that such stories happened, thus that the creative success of an agency became acceptable for big clients and their advertising performance criteria. As a result, the increasing importance of creativity inherent to the 'second wave' has strengthened the 'popular arts world' in a double sense:

First, creativity became a fashionable feature of agencies, visible, for instance, in the spectacular growth of creativity contests and in the emergence of a whole body of creative agencies often set up by the creative staff itself. In the 'first wave' the creative department in the agency still had the role of a production unit rather

---

39 Spieß and Schmidt (1997, p.138) quote an interview with Willi Schalk, former CEO of both BBDO and McCann-Erickson Germany: 'Creatives always have an internal perspective, that is, they orient their work to other campaigns [...]. That is also the reason for trends in advertising [...] where certain styles, mechanisms, philosophies are "in" for a certain time and dominate a whole series of campaigns.'

than a key function in the advertising process and the creative community celebrated their events in small and relatively closed circles.

Second, and closely linked to this shift to the favour of creativity, the barrier between creative reputation and business reputation has become more porous. Stimulated by the increasing ineffectiveness of a market research driven mass communication and, more importantly, by single success stories of creative advertising agencies at least single clients began to raise their awareness of creative reputation – however always influenced by the reservations outlined above. The example of the PR-effects of a *Cannes Lion* mentioned in the section on the second generation showed this shift regarding the perception of creativity. In terms of our model the thinner circulating arrow gains importance.

As a consequence of both processes the self-confidence of creative agencies has grown tremendously. Clients could be selected in accordance with their creative profiling given that creativity promised to provide economic success:

> I think, what distinguishes us is a very clear philosophy that creativity contributes to an efficient advertising. [...] And also our attitude is very clear in this context, that is, we do not make many compromises. [...] That means that we are by no means easy for our clients, we torture them a bit and from time to time we also stop the relationship if there is no way. One can of course hold that we can afford it since we are big and renowned now. Yet we already did it when we were small and actually needed the client. I think that this straightness in the attitude, of course along with professionalism and good work [...], that it does really matter in terms of the profile to the external world that in turn importantly pushes the fascination for both clients and staff (personal interview, 2000).

Also on the global level the predominance of British advertising characterizing the 'second wave' is based on this increasing permeability between business and popular art. The borders between the cultural and the economic in the UK had already been softer than in other countries before, for instance visible in the success that British advertising filmmakers like Alan Parker and others had in Hollywood in the 1980s (Lash and Urry, 1994). In Germany, for instance, advertising film for 'serious' directors still has the character of a '"secret lover": It is nice, but difficult to talk about in the public' (personal interview, 2000). Only recently there have been some notable exceptions even of famous filmmakers changing between the worlds.

It appears quite natural that the process of change also affected the structure of the agency landscape in that new and innovative agencies entered the market as depicted in the typology in Figure 4.1 (p.88). To a certain extent the separation between the popular arts world and the business service world has shifted from structuring the industry within the agencies, that is between account management and creative development, to structuring it between (new) creative and (old) marketing oriented agencies.

Yet, there is a second structuring logic when reading the model along the central axis. Thereby the economic or narrow monetary interdependence between the customer interface and the labour market interface is put to the fore. Again roughly put, an advertising agency can only survive if it earns money and it can only earn money if it has good personnel that in turn has to be paid by money from clients. This simple 'pecuniary' logic of the central axis is not without any impact on the relation between the two worlds. It even contrasts the actual friction between the cultural and the industrial logic of advertising. As creative reputation is achieved by successful participation in contests and as contests absorb agency resources, creative reputation presupposes literally an investment which often cannot be afforded by small enterprises with correspondingly low turnovers. This has been reinforced by the inflationary rise of creativity contests in the last years included in the creative rankings. Most of the big advertising agencies have in the meantime established own departments to handle the creativity business, as a sort of own advertising budget (personal interview, 2000).

In this logic the industry's second basic dichotomy of concentration on the global level and dispersion through the constant emergence of new agencies is shown, overlying the separation of advertising as business service and popular art. Global players, as they have to react to the new competition, use their financial power in order to penetrate the creative realm in three ways: First, they try to enhance the creative potential of their traditional agencies by attracting successful labour or the management of creative agencies. The BBDO-case quoted above (p.95) is typical for this logic. Second, they set up own creative boutiques which can directly compete with the small independent enterprises staffing them in the same way thereby having the advantage of a better image among creatives (Leslie, 1997). Third, direct acquisition of creative agencies leads to the same result, at the same time constituting efficient growth strategies to satisfy the shareholders of the transnational advertising groups (Leslie, 1995). In turn the formerly independent firms are provided with access to global markets and extensive facilities for market research, training etc., also their owners being properly rewarded financially (personal interview, 2000). Globally maybe the most prominent example has been the acquisition of Chiat/Day, the most notable creative agency in the USA of the 1980s (Leslie, 1997), by TBWA, a network under the umbrella of the Omnicom-holding in 1995 (Willenbrock, 2000).

However, this process of increasing concentration by itself opens spaces for the establishment of new enterprises, either by conflict with clients or by discontented management personnel which either themselves become founders or are likely to be attracted by a growing start-up. The fact that BBDO's buy-out of S&J management ended in the establishment of an independent agency is as illustrative as the Chiat/Day-example: With the take-over by the Omnicom-group, the staff of the London branch decided to leave the network, setting up their own completely employee-owned agency 'St. Luke's' which has been one of the most striking success stories of the last years in the world (ibid.).

Thus, put in terms of the model, it is the concurrence of the two axes that constantly restructures the agency landscape. The two dimensions of global vs. local and old vs. new which constituted the starting point of our case study analysis are expressed by a monetary vs. a cultural logic whose co-existence drives the continuous process of change. Put conceptually, 'second wave' advertising in fact corresponds to Lash and Urry's concept of culture industries as pioneer of economic reflexivity (Lash and Urry, 1994 p.123) in that it is basically structured and restructured through a dialectic of structure (as global advertising industry) and agency (as individual creative professional) in which 'agency is set free from structure [...]' by '[...] structural change itself' (ibid., p.5).[40] However the motives behind structure on the one hand and agency on the other remains different thereby implying a basic conflict between structure's desire to 'exploit' the self-reflexivity of the individual and the same individual's desire to be set free.

Regarding the landscape of advertising firms in Germany this dialectic at first glance depicts the duality of global players' increasing concentration and the continuing start-ups of the second generation. Two types of middle-sized agencies run the risk to be squeezed between these two dynamics: the traditional national players, which tend to react by specializing with their traditional client base as point of departure, and the creative pioneers. Yet, the latter play a particular role: First, they reflect the dual structure of the industry in the way they organize their agency as group of decentralized units. Thereby they try to manage the trade-off between growth and creativity (Leslie 1997, p.1033). Second, they are the base of the start-up dynamic by teaching their staff to become entrepreneurs. Nevertheless the limitation of the national market forces them to subordinate themselves at least partly to the financial logic of the international business.

That is to say, at second glance, the duality of concentration and dispersal cannot be understood as being directly driven by the conflict between global structure and individual agency but as fostered and mediated through the process of innovation that has taken place in the German advertising system boosted by the pioneers of the 'second wave'. Thus innovation did not only consist in having built bridges between the two worlds inherent to advertising but in having encouraged the emergence of a dynamic and 'entrepreneurial' labour market simultaneously needed in the global business and constituting an antipole to it. Before we shall examine the logic of this labour market in a more comprehensive and detailed way we shall briefly discuss the third dimension of variation within the German agency system: the role of location.

Proximity and Diversity: Advertising Agencies and the Logic of Place

In one of his papers on the London advertising industry Grabher (2001, p.355, 371) challenges the conceptualization of localities in the current work on learning firms and learning regions criticizing that it usually suggests a concept of learning focusing on the 'diffusion of "best practice" which, in turn, promotes organizational homogeneity and economic coherence' (ibid.). In contrast to this he argues that the 'Village' – as he labels the advertising cluster in 'Soho, the epicentre of the "second wave"' (ibid., p.352) – reveals a concentration of diverse organizational forms as well as diverse ways of understanding and carrying out similar activities. It is precisely this diversity lying at the heart of the continuous

---

40 See Chapter 2, p.31.

learning processes which makes Soho the most dynamic and successful location in contemporary advertising industry.

Our evidence focusing on the Hamburg case does not confirm these far-reaching interpretations without reservations. Of course imitation has taken place in the German advertising industry as regards the organizational structure of agencies. Springer & Jacoby's unit principle has been copied not only by the agencies split off from the pioneer but throughout nearly all agency types and throughout the country, thus not being a function of a certain locality but channelled through the industry as whole.

Nevertheless the clustering of a heterogeneous agency pattern both within a limited number of big cities and within and around the centre of these cities can be identified. The interregional concentration mainly affects the three most dynamic agency groups, that is, global players, pioneers and the second generation whereas both traditional national agencies and regional players seem to be more equally distributed throughout the German space-economy,[41] the former strongly oriented to the owner's preferences and the latter mainly driven by the existence of a potential client base. We shall take up this interregional dimension at the end of Chapter 5 in a more detailed way focusing on how the outlined process of innovation has affected the spatial pattern of the advertising industry in Germany.

The intra-regional concentration in inner-city areas can be identified throughout all categories, only the regional player tending to sub-urbanization due to rent market pressures and to the locational shifts of the client base. However even a huge part of them still shows a strong affinity to central sites ultimately motivated by the desire to use the environment and design of the office to present the agency to clients. This basic motivation to present the agency – its aesthetic quality, its business success etc. – through the office is certainly part of each agency's office strategy. However, besides this, the motivations that drive location tend to strongly vary with the agency type.

Whereas for global players the urban environment appears to have a very neutral and functional role, as a communicational node and a good setting for organizing the business in terms of supplier linkages, the labour force's desires etc., all other logics appear more closely linked to the specificity of each agency. In the case of the traditional players these basic functional aspects are strongly interwoven with the preferences of the owner, legal positions concerning property etc. In the case of the two new groups of players within the German version of the 'second wave' location involves much more: It is – of course in close association with its functional dimension – essential element of the overall profiling strategy of each firm according to the position it has in the course of its lifecycle. For the first generation this means that the office underpins the general philosophy or attitude of the agency, presently above all visible in the different understanding of

---

41 As an illustration: 62 of all German top 100 agencies (101 of the top 200) have their headquarters in the traditional advertising regions Frankfurt, Düsseldorf/Cologne, Hamburg. Out of the 61 agencies among the top 200 set up until 1980 and owned by a national majority only thirteen are based there (*Werben und Verkaufen*, 13/2001).

how to modernize it: with a more traditional approach by S&J and a more modernistic approach on the part of S&F, also reflected in the location. Unlike this, the young agencies of the second generation use their office and the location precisely as a means of distinguishing themselves from both 'classic', i.e. global network agencies and the 'prototypes' of the first generation.

That is to say, the clustering of different agency types in different locational environments of one inner city is not just an antithesis to 'best-practice' imitation but rather the concurrence of different locational logics. Generally it can be held that, with the rise of the 'second wave', agency location and the agency office has become actively included in the agencies' profiling strategies. As we have seen this profiling occurs on the client side as an element additional to classic reputation building. On the other hand, this would not explain clustering in one place since clients won't look for an agency by testing three in the same city centre. Thus, if this profiling within one agency cluster is a mechanism of 'fuel(ling) rivalry' (ibid., p.371) this rivalry might be the competition on the factor market, that is, for the best and most creative labour force that on one hand contributes to the profiling of the agency, yet on the other hand reacts itself to agency profiles in different ways. Thus, again, the labour market appears to be the key to understanding the spatial organization of the advertising industry. We shall now direct our attention to its functioning in order to grasp the use of this key.

Chapter 5

The Labour Perspective: The Creative Professional in the Restructuring of Advertising

The Creative Professional between Idiosyncrasies and Adaptation

THE COPYWRITER'S TEN COMMANDMENTS
[…]
3) Advertising is the only job in which you're paid for doing things badly. When you present the client with a brilliant idea and they want to 'make a few alterations', think long and hard about your salary, then cobble together the crap they're dictating in thirty seconds flat and chuck a few palm trees in on the storyboard so that you go and spend a week in Miami or Capetown for the filming.
[…]
7) Cultivate absenteeism, come to work at noon, never answer when people say hello, take three hours for lunch, and make sure no one can get hold of you on your extension. If anyone has a go at you about this, say: 'Copywriters don't work to a timetable, just to a deadline' […] (Beigbeder, 2002, p.39ff.).

As pointed out above it is its nature as a popular art which distinguishes advertising from other economic activities – of course not characterizing it solely but inherently interacting with its nature as a business service like consultancy, engineering etc. The key figure of the 'popular arts world' in advertising is the creative professional and, following our findings on the process of change advertising has undergone in the last about 20 years, this kind of 'worker' has played a major role within this restructuring.

Frédéric Beigbeder's story of Octave, a successful but disillusioned copywriter, depicts this creative professional in a rather sarcastic and above all stereotyped way; the quotations above do however touch two important aspects referring to arguments we have already used for the conceptualization of advertising as a whole, transferring them to the situation of the individual professional: The first one stresses the conflict between the ideas a copywriter develops, for instance as a script of a TV-commercial and the client's requests. Beigbeder's third 'commandment' proposes a sort of 'prostitution' on the part of the creative, accomplishing the client's wishes and sacrificing the artistic quality for his personal wealth. The second focuses on a certain 'peculiarity' that characterizes the creative in comparison with other activities (not only) in advertising. The seventh 'commandment' shows the protagonists as living in their own world of creatives, chiefly working on maintaining their image as eccentrics.

And indeed it is generally recognized, even in classic marketing literature, that the creative staff is special and does not fit the norms of traditional capitalist labour. As Wells et al. put it:

> The people who work for advertising agencies may not be at all like the employees of a corporation. Artists, writers, and television producers might not fit easily into the culture of the corporate environment. Rigid work hours, dress codes, and limitations on overtime would be difficult to enforce among the 'free spirits' who tend to work in advertising (1989, p.104f.).

Hence, a huge part of the labour force in advertising represents paradigmatic cases of the 'self-reflexive' individuals described by Lash and Urry (1994) or of Richard Florida's 'eccentric' 'bohemians' (2002), characterized by a high degree of self-determination. They are used to work on their own and to choose the environments in which they think to be creative.

However, it would certainly be a too romantic idea of the creative staff in advertising to portray it as entirely driven by high artistic demands and therefore orienting its decisions in terms of work environment etc. exclusively to the fulfilment of these standards. Seen in this way such artists would appear immune against every kind of pecuniary temptation: Thus, they could on the one hand easily be exploited by capitalists in terms of cost, their wages being kept down without any resistance. On the other hand it would be difficult to make their work adapted to the needs of advertising, and this independent from the wage they are offered.

Given that advertising basically consists in 'selling ideas' it is part of the very nature of creative labour in advertising to be confronted with the business service world; put in more general terms, it is the core task of advertising to handle the conflict between artistic and monetary demands. Beigbeder cynically suggests one specific way to do this, in terms of artistic standards surrendering to the necessities inherent to the client sphere.

We hold that, despite the obvious roughness of Beigbeder's portrayal, the two aspects he outlines, that is the idiosyncrasies of creative labour and its necessary confrontation with and partly adaptation to the business needs, constitute the basic ambivalence that shapes the logic of the labour market of creative advertising professionals, at the same time providing excellent points of departure to understand this logic including its changing position within the functioning of the whole industry in the course of the last 25 years. We shall now briefly outline the contours of this change from the labour standpoint and then use two example of creatives' professional biographies in order to focus on its two specific features, that is, the high degree of inter-firm mobility and the importance of personalized relations within the wider professional milieu.

Creative Labour in the Innovation Process: Access and Adaptation

We have argued above that the innovation boosted by the main players in German second wave advertising had basically two dimensions: first, the bridging between the different conventional worlds of advertising combined with a creative turn, respectively with an increase in value on the part of creativity standards, and, second, the generation of a dynamic labour market constituting an antipole to the pervasive globalization process the advertising industry has been undergoing. Translated to the functioning of this labour market these two dimensions appear to be closely interlinked.

As regards the logic of increasing permeability at first glance the innovation corresponds to what Lash and Urry describe as main process innovation of the second wave, that is, it succeeded 'to reconcile the creative and marketing research approaches' to advertising through the introduction of account planning, being 'emblematic of the implosion of the economic, advertising as a business service, into the cultural, advertising as a "communications" or a "culture" industry' (Lash and Urry, 1994, p.141). Even if this change can less be considered as an institutionalization of account planning as 'third column' of the advertising agencyit appears justifiable to hold that the two key 'job descriptions' of advertising, account management and creative development, have to a certain extent tended to converge, or at least to come closer together, the bridge between 'selling ideas' and 'creating ideas' thus having been built inside the agency.

Konstantin Jacoby and Jean Remy von Matt, probably the creative protagonists of 'second wave' advertising in Germany, consider this convergence in different ways. The former sees the new relation as a sort of return of the traditional advertising consultant and as a stronger personalization of advertising: 'In the 1970s and 1980s clients mainly trusted in agency organization and agency systems. [...] Meanwhile they ask: "Where is the beef? Where are the people that help to solve my problems?"' (Jacoby, 1995, p.126). This view appears surprising given the inherently organizational character of the Springer & Jacoby strategy. Unlike that von Matt more strongly focuses on the original S&J-culture, that is on the internal atmosphere and the degree of commonness of an agency: 'Ultimately there are two types of agencies: those whose account managers and creatives never have lunch together, and those whose account managers and creatives often have lunch together' (von Matt, 1995, p.149).

In addition, symbolizing the increase in value of creativity, the creative development functions (and thus, jobs) within an advertising agency that is, copywriter and art director, have come to play a much more vital role within the process of building advertising campaigns, by being involved already in strategy development stronger than they had been before. For instance, creatives' bottom-up marketing strategies based on and built around the 'big idea' have become possible options to start successful campaigns (Schmidt and Spieß, 1994, p.64).

But the changes have not only affected the internal functioning of advertising agencies but have strongly interacted with the labour market system as a whole, in several respects: First, the labour market of creative professionals has grown in quantity. As could be seen in Chapter 3 not only the sector as a whole has increased its employment, but particularly creative tasks have expanded in comparison with management and clerical activities.[1] Yet this is not only a matter of the demand side on the labour market; the rise of creative activities is closely linked to the increasing participation of women in the labour market. In our regional case of Hamburg, for instance, the share of women in creative activities – not only in advertising – increased from 33 per cent in 1980 to 46 per cent in 1997 (Läpple and Kempf, 2001b).

Second, jobs in advertising have generally become more attractive. Asked about the image of advertising in the middle of the 1990s, again Konstantin Jacoby and Jean-Remy von Matt illustrate this change:

> When I began with advertising, having just concluded my studies in 1975, all my colleagues had the dream to get a job at one of the big weekly journals: 'Spiegel', 'Stern', or, even better, 'Die Zeit'. I had studied Communication and Germanic Studies, and advertising was rather considered as a personal anticlimax. Today you probably have a certain status when you get a job in an advertising agency (Jacoby, 1995, p.126).

> It is interesting that journalist and advertising professional to a certain extent have exchanged their images. Whereas the journalist has lost credibility in the past decade, having in the struggle for a high circulation turned from an objective reporter into a sensationalist man hunter the advertising professional has added to his credit. Formerly having been the 'hidden persuader' that roused desires actually nobody had, he is today a positively thinking person aiming at clarity and bound to truth, who brings experiences like Toyota animals and the Aral walker into our living rooms (von Matt, 1995 p.148).

Thus, whereas jobs in the advertising industry of the 1960s and 1970s still suffered from the stigma of a 'morally low' (Jackson and Taylor, 1996) activity, they now appear to be worth striving for.

The roots of this image change of advertising professions are difficult to identify. Schmidt and Spieß (1994), drawing on interviews with 32 creatives, ascribe it to the higher degree of acceptance of advertising as a whole through the fact that the generation entering the labour market in the 1990s had grown up with advertising as part of its everyday culture. In addition to focusing on changing values in the potential labour supply, gradually 'accustoming' the society to accepting advertising as both part of everyday life and tolerable activity, we hold that it was also through changes in labour demand, that is within the very functioning of advertising, that it became generally more attractive. Given that creative skills were getting higher importance within the advertising process the potential labour force saw in it new possibilities to realize its high demands in terms of both work quality and work environment. The labour market of 'second

---

1   See Tables 3.4 and 3.5.

wave' advertising thus emerged as a circular process, in which creative advertising needed creative staff that on its part could be attracted by creative advertising.

Third, the key to setting this circular logic in motion automatically was to get access to labour market segments previously standing outside the logic of rigid economic exploitation, namely visual arts, literature, journalism or even fields more distant from business. Particularly the function of the copywriter changed with an advertising strongly based on 'story-telling', particularly given the increasing importance of 'narrative' TV-commercials, and needed really new people.[2] This access can be understood in two different perspectives: On the one hand these segments had to be convinced to integrate them into the capitalist logic of a business service. In the case of Springer & Jacoby this occurred according to the pattern of a cultural innovation and innovation-diffusion (Wolfe, 1999) through presenting the agency as something unique and completely new and different from 'normal' advertising, thereby first rising from being a 'personal tip' to a special school and community and then to a renewed image of the whole industry. On the other hand advertising offered new opportunities for people previously little inclined to private sector occupation. An art director explains her difficulties in looking for career opportunities corresponding to her talents and interests before finding the way into advertising:

> I had always concentrated on creative things as a whole. I have never been interested in science or all this dry business stuff. And, once being at a vocational adviser, I really cried since he recommended me to become china painter. And as I then expressed that I was interested in arts and literature, he replied that there would not be anything, that it was a pity that I was not interested in computers since with what I desired I would not be able to earn money. Either china painting or normal vocational training in a bookshop were the alternatives he suggested. In the end advertising as another alternative worked quite well (personal interview, 2001).

Both perspectives, that is, seeing advertising as 'hostile' business world or as opportunity to earn money with artistic talents have been certainly strongly interwoven in the opening of the labour market segments of creatives and may differ from individual to individual. Over time the opportunity-view may have outweighed the sceptical position due to the improving image of the activity. Nevertheless a stronger involvement of people with distance to business to a certain extent reinforced the basic conflict inherent to the labour market of creatives, between the personnel's specific demands and the need to adapt their creativity to the demands coming from the client sphere.

Thus, coming back to the two key dimensions of the innovation process, as regards the generation of a dynamic labour market, the innovation involved a

---

2   The copywriter is still an occupation with a largely undetermined professional profile. One managing director of a big agency reported on their good experiences in employing theologians and educationists as copywriters. 'They are used to transmit messages which are not their own' (personal interview, 2000).

double 'bridging' of business world and popular arts world: On the one hand, the labour market segments of 'artists' were opened and provided with an important role in the development of advertising strategies, having in mind the peculiarities of the creative people. On the other, their creativity had to be integrated into strategies which cope with the needs of advertising, or, in other words, it had to be tamed down to a certain extent, according to the restrictions set by the 'materiality' of the business (Paczesny, 1988) in terms of formats, marketing requirements etc.

In the case of the key pioneer Springer & Jacoby two strategies have been crucial to handle this tension, more or less having the character of a bargain between the agency and its employee. The first strategy is training. In terms of the bargain this means that the creative employee declares himself willing to adapt to the needs of advertising or, even more importantly, to the basic laws of the agency which comprise – with 'client, cash, culture and creativity' – all parts of the advertising business. In turn, as second strategy, the agency offers to treat him or her as individually as possible, as regards the tasks he or she accomplishes, the possibilities to promote etc. Yet, this offer is not made for altruistic reasons:

> You have here the possibility to develop in the sense that you some day recognize: 'That's what I actually do best.' And the principle behind this is: 'Put the people where they can develop the most enthusiasm, since there the best things will happen' (personal interview, 2000).

In addition the agency supports each employee in fulfilling his or her demands as to creativity standards by keenly promoting success in creativity contests. That is to say, besides providing each employee with the possibility to 'get as far as he or she wants to' (personal interview, 2000) Springer & Jacoby accomplished the wishes to earn reputation in the community of creatives. That is to say, the second dimension of innovation was not only based on the double 'bridging' of the two distinct systems of reputation building but essentially required that artists could become advertising professionals without giving up their creative or artistic identities and capacities.

In order to fully understand the ambivalence of access to artistic labour market segments and their adaptation to the needs of the advertising business three further aspects have to be taken into account: First, the possibility to maintain an identity as artist neither automatically nor completely abolishes the basic contradiction between the two worlds of reputation building and the different 'performance criteria' they are based on. It still has a sense of compensation since the actual substance of the creative turn continues to consist in 'the formation of more sophisticated advertising techniques for mediating resistance to consumerism' (Leslie, 1997, p.1021) and not in the fulfilment of artistic standards. Second, the dialectic of inclusion and separation regarding the professional identity has a different character when looking at the process of creative development and production. Although it is clear that 'many creators have an ambivalent relationship with the business of creativity' (Bilton and Leary, 2002, p.56) the existence of contradictions and frictions is generally considered to constitute a positive stimulus for creativity (ibid.; Grabher, 2002). This also applies to the 'materiality' of the business, such as deadlines, format etc., given that it provides a clear framework (or 'structure') in which creative action can unfold (Nov and Jones, 2003). Third, of course the labour market innovation interacts with the changing agency landscape discussed in Chapter 3, concerning both the emergence of new players and the relative weakness of the traditional ones and the fact that global networks tend to compensate their weakness by 'investing' literally in creativity.

We shall in the next section try to obtain a deeper insight into the functioning of the labour market of creatives, discussing it as the concurrence of a changing agency landscape and more 'reflexive' individuals simultaneously produced, needed and affected by it. The discussion begins with depicting the professional biographies of two art directors, thereby taking up a methodological and didactical tool frequently used in the recent analysis of changing career paths and work patterns. Richard Sennett's Rico (Sennett, 1998) and Florida's examples of a 'life in the horizontal labour market' (Florida, 2002, p.113) stand for the increasing destabilization of biographies, respectively for individuals' increasingly active 'navigation' through their careers. The following 'two stories' are less spectacular than the models provided by Sennett and Florida; rather they attempt to illustrate how the mainstream activity in advertising constantly implies to deal with the intricate interaction of own desires regarding the making of a (not only) professional biography and the economic structure which constitutes both its precondition and constraining framework. The key characteristics of the labour market subsequently discussed constitute so to speak the obvious outcome of this interaction and of the risks and uncertainties it implies.

Life in the Creative Labour Market: Mobility and Personalized Relations

Two Stories

Andrea's[3] artistic career already started at school. She deepened arts in the upper secondary school and, in addition to the normal lessons, she loved to work at the student's journal, being responsible for the layout. After her school-leaving examination she began a course in Germanic and Media Studies at the university, yet at the same time applying for a three years course at a graphics school in Berlin. She was accepted and finally liked the course for learning the basic skills of graphics, typology etc. even if she later noticed that it did not much deal with the later practical work, chiefly in terms of the time rhythm of projects.

Andrea finished this course as scheduled and soon got a job as volunteer in a small advertising agency where she worked with an art director, doing small things

---

3   Original name changed.

such as designing a flyer for a prisoners' aid organisation. What she drew from this first experience in an agency was that the people in advertising were interesting and that the main job consisted in dealing with one's imagination. Her art director, for instance, was a trained painter and photographer, working as graphics professional in the agency. This first practical thus could demonstrate that a career in advertising should be her way to go.

Following to that she applied for a further practical advertised in an advertising journal, in the branch of a renowned network agency in Hamburg. Hence Andrea moved to Hamburg for three months and both the city and the work in the agency were fantastic, showing her, after having worked in a small and relatively insignificant agency, how advertising could be done in a professional way. Nevertheless, after these months she returned to Berlin for a job in a relatively big agency, however keeping in mind that she would later like to change to Hamburg. Now she had become junior art director, that is, she developed small things within a campaign in close co-operation with the art director; thus she was not involved in the client contact nor did she collaborate directly with the copywriter. After three years this relatively limited set of tasks no longer satisfied her ambitions and she looked for a possibility to advance, finally applying for another junior position, but now in a big Hamburg based agency.

Here she was lucky given that the art director left after one and a half years and she could take his place. So Andrea suddenly had the responsibility for an account and noticed that it worked: that she was able to present ideas and to defend them, that she had learned to select photographers together with the art buyer, to prepare presentations for new businesses etc. However after some time the agency was sold and the management changed completely. With the old management's departure many of the colleagues left, some clients got lost and the atmosphere within the agency substantially degenerated. When the client Andrea was working for also renounced, she decided to leave without having an alternative and spent the next months with freelance work, mainly for small agencies and with jobs that the in-house staff evidently did not want to do. In addition she supported her boy friend who was setting up his own business at that time, producing his firm presentation folder and doing other small things for him.

But, given that she worked more or less on her own, she missed the professionalism of a big agency with specialists for jobs like art buying and broadcast and print production, with team orientation and with more financial resources which for instance allow to buy more expensive photographs etc. Thus she looked for a job again and finally got a call by her former creative director who at that time was working in the branch office of an international network agency. Despite having actually planned to look for something more 'sparkling' in terms of creativity Andrea accepted because she knew from the former collaboration that she liked this colleague and expected that at least the atmosphere would therefore be good.

She did not regret this decision since she could then work well in a good team, for the first time with her own copywriter with whom she got along, working for good clients, and she enjoyed having left freelancing. However there were again changes propelled by the national headquarter in terms of account structure and team composition that spoiled the atmosphere, again driving many people to leave. Andrea also left applying for the fourth time in her life at one of the two pioneer agencies without any job advertisement. She was invited to an interview in the team she had actually applied for; as this team recognized her qualifications but did not need her profile, her papers were given to another one where she finally was employed as art director. This job gave her for the first time in her career the opportunity to work in a bigger team responsible for a whole and big cigarette account, from small promotions to cinema commercials, including participation in world-wide photo shootings, having her 'own' junior and volunteer needing to be instructed etc.

Andrea has been now in this agency for two years. She is now 36 years old, that is, relatively advanced for an art director, thus being aware that she both has to look for further promotion, that is, a job as creative director, or for a way out of the narrow creative work. Except being sure that she won't end in her present job she still does not know what the future holds for her.

Mary[4] had originally intended to become industrial designer through a combination of a craft vocational training and a university course. But her parents thought that she had to begin with a university course given that she had obtained the right to higher education. So she surrendered and began a course at a private graphics school in Hamburg. As the quality did not meet her expectations she already changed school after six months, her second attempt being successful. She finished this school as scheduled and was directly employed as junior art director in the agency in which one of her teachers worked. It was an old traditional Hamburg based advertising agency, at that time being affected by a significant drain of good people. When also her art director left to a spin-off of this agency after six months she followed him.

However, Mary noticed very soon that this decision had been wrong, since the atmosphere in the new agency was terrible, leading to the highest labour turnover rates among all advertising firms in Hamburg, the creative part having been completely exchanged within two years. People were dismissed without previous notice. One of the victims of this crude labour strategy was the art director she had followed and she was also working for. At first she took his place in being solely responsible for the account but after some time a new art director was employed. After two and a half years she finally left the agency and switched to a small five people firm the contact to which she had got through one of the management members. The job there was very nice given that Mary could acquire experience in all possible fields of the creative process, from development to final layout and production; she was even involved in account management. Again after some time the agency should merge precisely with that one Mary had left before. She actually accepted to stay due to the loyalty to her boss at that time, notwithstanding the

---

4   Ibid.

obvious reservations given the experience she had made in the agency before. But during holidays – having already signed the new contract – she got a call from the big agency she has now been working for for about two years. Her first art director whom Mary already accompanied when leaving her first job is now the creative director of the group she is working with.

Labour Mobility: Between the Individual and the Industry

Andrea's twelve-year career as graphics professional in advertising has confronted her with seven different jobs if counting her freelance intermezzo as one. Also Mary works already in the fourth agency after having finished her graphics course although this was only about seven years ago. The two examples mirror an enormous dynamic which appears to drive professionals from agency to agency, respectively from agency to periods of self-employment and vice-versa. This dynamic seems to be a key characteristic of the labour market of creative professionals. Experts estimate that on average an advertising agency renews one third of its creative staff within one year (personal interview, 2000). Another typical statement is that creatives typically change their job every two years (personal interview, 2000). In this sense our two art directors more or less represent the average of the labour market.

The two biographies have also shown that the reasons for the extreme turnover rates in the labour market are complex and that they can neither be completely ascribed to the idiosyncrasies of the labour force nor to a wider structural logic of the business. As held above they display again how structural and individual aspects interact in the functioning of contemporary advertising.

From the individual employee's perspective inter-agency mobility on a very general level originates from dissatisfaction with a given work situation. A present job is not able to fulfil the individual demands regarding three interrelated key aspects of the work life, that is first, the *strategies of, respectively the desire for individual advancement*, second, the *nature of the work* itself and, third, the *quality of the work environment* in terms of collaboration, management etc. If this dissatisfaction cannot be corrected within the borders of a firm an employee is likely to change.

Yet also the three aspects are by no means completely consistent in themselves. *Individual advancement* is first of all professional upgrading in terms of doing more complex tasks, assuming more responsibility, but also working in a more professional firm with more demanding clients, a more sophisticated division of labour within an agency etc. Andrea's career is strongly characterized by this desire to advance in professional terms. Particularly her return from self employment into agency work, but also most of her motivations to change show this aspiration to do good and professional work, collaborating with other professionals in production, art buying and, above all, copywriting. Also her final move to Hamburg was driven by the former experience she had made as volunteer in a more professionally organized agency, that a close collaboration between art and copy already on the level of volunteers contributed to better results. In this context, professional progress does not necessarily imply the path from small to big agencies:

> Sometimes you note that you should work now in a smaller agency where you are forced to assume more responsibility for certain tasks or where you have to do a greater variety of things, not only one specialized task as in a big agency (personal interview, 2001).

In addition, the personal progress depends on the work one has done before, above all in terms of creativity standards, given that applications are generally based on the personal work portfolio drawing precisely on the actual work:

> I had to make this experience. As I had spent too much time in middle class agencies I did not have anything outstanding in my portfolio. And it was relatively late that I thought: 'Now I want to go to S&J or so', and my portfolio was really a barrier. And I was torn out of my sleep for the first time and thought: 'Now I have to work for my portfolio.' And if you cannot draw on your job for it you have to do this in extra-work beside your job. It is expected: If you want to have a fantastic job, you have to work for it (personal interview, 2001).

Thus advancement obeys a set of industry-specific rules of how labour force is 'priced' derived from the canon of arts: one's capacity to cope with the demands in artistic terms proved by a sample of previous work. The fact that this sample cannot be drawn from the present work may in turn also be a motive to look for something new.

In addition, advancement does not necessarily occur within the narrow field of classic advertising activities. Both examples also show that the way into advertising has absolutely not been automatic. The first experience of practical work functions as a sort of try-out period whose end is actually open being likely to lead to jobs in other fields:

> There are, for instance, people who have become junior art director but who notice then that they do not intend to become art director but concentrate on, say, design. And then they look for agencies specialized in design or so, and they leave (personal interview, 2000).

And, the advancement can also have a strong orientation towards climbing the ladder in terms of one's career:

> Some people are very strongly focused on their careers. They already try to start in the best agencies – at any cost – and they look for a post as creative director as soon as possible. I have always concentrated on working with the people I like, on having the opportunity to learn a lot and not on getting a CD-job, since as CD you do not really work any more but are strongly involved in agency policy etc. (personal interview, 2001).

The career path does not necessarily mean to promote within the occupational hierarchy. Creatives may make good jobs but are likely to be inappropriate in management positions. Agencies thus sometimes even prefer to keep particularly copywriters in lower position but paying the same wage as a creative director (Hattemer, 1995).

In sum each creative's objective to obtain a certain degree of professionalism interacts with the search for an individual professional identity, be it in terms of specialization or in terms of an attitude towards the individual career. Employees tend to change the agency if they do not see the possibility to advance both professionally and personally in their present job.

As regards the *nature of work* in an advertising agency the satisfaction of a creative grows with the extent to which personal creative demands can be met. It is, roughly said, the creative's job to continuously develop ideas. The actual work in agencies is yet generally organised as one creative team working for one account.

> We had a creative who had worked ten years on a petrol account. And then he could change to a small regional cheese label. And his work was excellent. He was highly creative since had yearned for something new (personal interview, 2000).

Thus the concrete job one does in the daily work may or may not accomplish the individual ambitions in terms of creative standards, be it due to the character of the client or simply due to the fact that after a certain time ideas for always the same product tend to be exhausted, the ten years shown in the quotation already being an extremely long period. Put in terms of labour mobility: As exhaustion of the creative capacity in relation to one particular project cannot always be avoided within the firm, employees change, so to speak to get 'fresh air'.

A similar logic is at work as regards the *logic of the work environment*. Ideas do not rise simply through the capacities of an individual but they are supported through the context in which it is embedded in:

> We always began with a first meeting in which we make a brainstorming, that is, everybody can express what he just thinks about in terms of the product and an eventual campaign, even if it is really nonsense. And perhaps this nonsense may inspire somebody to make a real idea of it. And you can have this nonsense from a film you have eventually seen the day before or from a comic you have read the day before, or a book you have read in your childhood (personal interview, 2001).

The advertising work is thus fully teamwork and the results strongly depend on the functioning of this team, beginning with the close co-operation between art director and copywriter but also comprising the wider context of juniors, volunteers, producers etc. To work in a well-functioning team is one of the main ideas creatives engage in. In turn, a team that does not work well is one of the main motivations to change the agency. Both biographies outlined above show that changes in the narrow social setting of daily work always constituted serious 'push factors', even once making Andrea temporarily prefer a more precarious situation as freelancer.

> The new management was really strange, the atmosphere continuously worsened, many good and nice people were leaving the agency, and others followed them. I was really having the feeling that one day I would be last. At last, even my jobs lessened, I actually had almost no more work. And then I renounced without having something new. And I was really sad and thought that I would never have the chance to return into an advertising job (personal interview, 2001).

> Between me and my copywriter it is sometimes like being married. Well my copywriter is a really funny nice guy, really nice. And he recently renounced being enticed by another agency. And I was sitting in my office crying one hour, I thought someone would have torn out one of my legs. [...] Fortunately he will stay now. All the colleagues urged him to stay and we succeeded in convincing him (personal interview, 2001).

Thus, the quality of interpersonal relations appears to be of major importance for the individual career and for the decision where a creative professional likes to work. From an individual's perspective hence the high mobility rates are obviously owing to creatives' high demands or standards in terms of their work in all respects which in turn is yet necessary as regards the nature of this work:

> Among creatives mobility was always higher than in all other fields, as long as I can think back. And this is not without reasons, since to be creative means to be more sensitive than account managers. The latter need to be thick-skinned, given that they have to deal with customers. Yet the former need to be sensitive to generate new ideas. And so they have so to say a shorter life span in their job, they more rapidly tend to be burnt out. Basically they have to produce more with their head, their intuition, actually out of themselves (personal interview, 2000).

From the perspective of the advertising system there are four major aspects which foster inter-agency mobility, affecting different logical levels, from the nature of the narrow labour process to the global concentration of the industry, that is, the *labour intensity* of advertising, the *nature of the creative process*, the *scarcity of good workforce* in the face of the basic conflict between the financial and the cultural logic of the industry and the *volatility in the field of management* caused by the continuous restructuring of ownership patterns.

As regards *labour intensity* one can hold that an agency's labour demand reacts to changes on the client market in a very elastic way. Given that at least from a certain size of regular customers' accounts there has to be one team working exclusively for it, both winning and losing a new client in general implies that the employment level goes up, respectively decreases. In the case of growth this means that agencies with each new client have to look for corresponding labour force. In the case of decline labour market regulations may hinder agencies to directly react to these changes in the sense of immediately firing the people working on affected accounts. Yet, as the quotation above shows, the fact of not having any real task frequently drives employees to voluntarily leave the agency.

In terms of the *creative process*, the 'need for fresh air' is not only a desire coming from the creative staff but being inherent to the advertising process respectively to the business strategies of agencies. Creative work towards advertising campaigns which mainly aim to attract consumers' attention by distinguishing the promoted brand from others needs surprising elements. Too fixed environments tend to undermine a surprising creativity, or, argued the other way: As creative work is solely embodied in the individuals who do it, changing constellations of individuals are likely to produce changing outputs. From an agency's perspective this need for flexibility in putting teams together has yet to be balanced with a coherent enterprise culture which is necessary for branding the enterprise in the different spheres and creating a platform to motivate people. Thus the management of staff in advertising agencies is driven by a need for both stability represented by a strong core of 'carriers of the agency culture' and flexibility represented by a 'healthy' rate of labour turnover (personal interview, 2000), the latter even making creative directors or other management representatives recommend staff to look for something new.

Concerning *scarcity of labour* the strong competition for the best people implies that enticing employees away from their agencies is a common habit in advertising. Two opposite logics can be observed here: On the one hand, big network agencies try to attract creative staff by offering higher salaries, stock options and an international work environment. In this context again the creativity contests play a crucial role since the decoration of employees for successful campaigns raises their values on the labour market and the likelihood to be courted by the big players in the business. On the other hand there is also the way back, that is, small and particularly fast growing agencies try to attract people in order to cope with their speed of growth. Experienced people from big agencies are very welcome for those 'second generation' unable to wait for the fruits of their own training and promotion strategies. Here it is less pecuniary pull factors which make people change their jobs but rather push factors of discontent with work in a corporation-like transnational advertising network.

As to the *volatility of management* both concentration on the global level and buy-outs of small agencies in the national market by global players affect the management level of local agencies, respectively the local offices of international networks. Even more, both ways of reorganization tend to impinge on the complete staff of an agency. In the case of changes within a global network the main driving force are the targets set by the global headquarter all members of a holding are bound to. When failing these targets a local management is likely to be replaced, leading to the replacement of lower ranks as well since the new management tends both to bring in new people through which they want to reduce the risk of failure in terms of the challenge to improve the agency's performance and to 'expel' a part of the existing staff that flees from the changing atmosphere within the agency emerging through the new people. To give an example of the impact of a management change:

> When we got, last summer, a new Chief Creative Officer for the whole German group, this triggered a nearly complete change of the creative staff in two agencies, that is, here and at XXX where she came from. Thus our business, and particularly the part of creative development is extremely based on personal relations, affinities etc. (personal interview, 2000).

In the case of local buy-outs the very management changes are caused in other ways; the impact on the internal labour market however tends to be the same. The biographies of Andrea and Mary both show various cases in which management changes lowered the quality of interpersonal relations within the agency thereby making employees eager to leave.

In sum, the volatility of the labour market of creatives in advertising is driven by the different facets of the business which interact and are thereby reinforced mutually: the staff's high demands in terms of labour constellation, individual tasks and personal identity building, the economic features of a business service, the nature of the creative process and the conflicts inherent to the industry between its cultural and its economic logic. As we could see all this is both strongly underpinned and shaped by a very personalized type of relations which channel the individual creative worker through the overall system throughout his or her professional biography, which however also provide agencies with an orientation within the labour market helping them to reduce the uncertainties inherent to its volatile nature.[5]

Personalized Relations: Mutual 'Investment in Contacts'

At first glance it would be easy to argue that the personalized character of the relations in the advertising labour market is mainly owing to the idiosyncrasies of creative labour. The sensitiveness necessary for their job, the fact that they basically 'produce out of themselves' also implies a more 'emotional' way of interpreting interpersonal relations, so to speak a strong affectivity which does not only have effects in the private realm but strongly influences the professional life as well. Even if this argument is legitimate to describe and to characterize the relations it lacks in neglecting the inherently social character of the advertising business.

Towards the end of Chapter 4 we held that social relations and institutions are the more important for the functioning of advertising, the more complex the business is, complexity in this context referring to the innovation-orientation of the campaigns, market size etc. Concerning the labour market we ascribed this social underpinning of market relations to the fragmentation of the market and to the

---

5   See the discussion on the nature of relations in the advertising industry at the end of Chapter 4, pp.110ff.

nature of labour as a 'fictitious commodity'. Given that creative work is solely embodied in the single employee this nature is carried to an extreme since at the moment of engagement it is still highly uncertain whether an employee's performance really meets the expectations. In addition, as we could see, this performance does not only depend on the single employee but is driven through a combination of his or her creative capacities and the relational constellations the creative is involved in.

Granovetter (1992) pushes another aspect to the fore, assuming a causal relation between high turnover rates and high rates of 'contacts', as he very neutrally labels the personal relations, seen from the perspective of an employee:

> As one moves through a sequence of jobs, one acquires not only human capital but also, and more difficult to interpret as an investment phenomenon, a series of co-workers who necessarily become aware of one's abilities and personality. This awareness occurs without cost as a by-product of the interactions necessary for work; [...]. Because of the often-documented fact that employers acquire a great deal of information about prospective employees from individuals known to both, one's market situation changes significantly with the number of individuals who know one's characteristics and with the number of firms in which they are located (ibid., p.239).

That is to say, the wider an individual network of professional contacts is, the better is the market position of an employee. A reservoir of contacts is the base for getting ahead via 'lateral' (Lash and Urry, 1994, p.200) or 'horizontal mobility' (Florida, 2002, p.104). This relation between mobility and contacts can be applied to the advertising labour market without restrictions, as the following remark shows:

> That is actually a sort of incestuous 'cliquishness': You go where people are that you know, and these people give you a hint as to somebody else who has heard about somebody etc... And so you move through a diversity of agencies (personal interview, 2001).

Yet it is not only the quantity of contacts that matters but also aspects of quality, in different respects: On the one hand, the contacts presuppose a good common work experience in order to really improve the market position. The virtuous circle of contacts and chances can also be a vicious circle of information about failure circulated through the professional milieu and worsening the individual market position with every additional contact. On the other hand, in the case of creative workers it is not only – and above all not in all cases – the improvement of one's market position that drives the individual 'investment in contacts' (Granovetter 1992, p.256). Trustworthy contacts do also reduce an employee's risk to work in environments that do not correspond to the personal quality expectations in terms of both professionalism and atmosphere. Both Andrea and Mary within their careers have even preferred agencies to which they had reliable contacts to those actually more attractive in terms of reputation and career-building. We have shown above that individual employees differ concerning their attitude towards the career. Yet, in what they do not differ is the very fact that they invest in contacts strategically in order to both reduce the risks of a volatile business and to guarantee the fulfilment of their individual objectives.

From the perspective of an agency one can also describe labour strategies as 'investment in contacts'. Given the double risk of an undetermined actual performance of a creative in an untested interpersonal constellation within an agency a pool of trust-intensive contacts in the labour market may reduce the uncertainties inherent to the complex process of creative labour. To know somebody who knows the capabilities and personality of an eventual candidate, and to know that one can trust his or her judgement, is an invaluable resource for an agency. It is even more important since the volatility of the overall labour market shown above requires rapid reactions to changes either through abandonment on the labour supply side or through a new account which exceeds the present labour capacity on the demand side. As we have just shown certificates, but more importantly work portfolios and copy tests, play a major role in that they reduce the uncertainty according to professional standards thereby limiting the number of potential candidates to be recruited; they do however not guarantee that these professional capacities can also be unfolded in the given environment of the agency. Trustworthy contacts of course cannot ensure success; they yet can reduce the uncertainty in terms of the interpersonal character of creative work.

It is another important element of the innovation process within German advertising that the two new agency generations have professionalized and even 'industrialized' the co-ordination of this 'contact pool' by establishing the corresponding infrastructure within their firms:

> We dispose of a very big database of 13,000 people. Of course we do not manage to observe every step in the career of each, but in principle this is our task. And there is a whole body of 'flops' among them, no doubt [...]. Well we have 3,000 contacts every year and among these 3,000 there are at least 500 really top people that I would like to employ immediately. Of course you cannot employ them all, but we try to keep in touch with them and to observe where they are at the moment etc. (personal interview, 2000).

The agencies' enduring success in maintaining their attractiveness for creative labour underlines the importance of actively 'investing' in personal relations in order to cope with the nature and functioning of the advertising labour process and market. Granovetter thus, although being clearly right in that there is a causal and interactive relation between turnover and personal contacts, does not cover all aspects of the advertising labour market when limiting this relation to individual promotion strategies of employees. 'Investment in contacts' is a mutual phenomenon which helps to improve market positions and to reduce risks for both workers and firms within the business.

Dealing with the Limits: Working Time and Age

There is an additional, rather cynical explanation of the high degree of personalized relations within the advertising labour market derived from the high personal strain professionals are confronted with based on the working time patterns of the industry: Given that the employees quasi never finish their work they have no other chance to build affective relations than within the professional context. And indeed overtime work at night and at the weekend is normal and the nature of relations within the labour market in fact appears to partly blur the borders between the spheres of production and reproduction.

However it would be seriously misleading to discuss these working time patterns merely as a means for labour exploitation through the agency capitalists, even if they do constitute fundamental restrictions for particular groups of the labour force, above all for women who try to balance their job with obligations of childcare. Capitalist exploitation does not correspond with the active and activating human resources strategies inherent to the business as well as it does not fit the very strategic behaviour single professionals reveal on the labour market. If thus exploitation shall be the right notion to talk about creative work in advertising a voluntary self-exploitation appears to be the more appropriate description.[6] As could be seen, creative professionals are extremely demanding as regards their work results, work environment etc. since they aim at identifying themselves with what they do. Once having succeeded in meeting their demands strong identification implies strong commitment and responsibility for the tasks one has to do.

Personal relations within the professional milieu in this sense function as a 'community of equals' (Piore, 1990) through which the individual professional identity (or the personal biographical narrative) is constructed.[7] Only through having a community which is able to assess one's capacities in a professional way this identity can really unfold. Yet, in relation to the working time patterns, the personal ties within the wider milieu play the further, less symbolic but more material role of averting risks inherent to the 'uncompromising' nature of advertising work. Taking up the example of women with small children, they may provide possibilities of both temporary freelance work, according to the time resources left by the family obligations, and of a re-entry into regular employment. Moreover, a high number of trustworthy contacts may even help to generate opportunities for the time after creative advertising work in a narrow sense, given that the biographical period of creation ends relatively early. Andrea, still being art

---

6   There is an intense discussion on new patterns of work and new labour force oriented firm cultures and their critical impacts on the 'work-life-balance' of employees which would exceed the limits of this book. As a benchmark study see Hochschild, 1997.

7   In organization studies the term 'communities of practice' (Brown and Duguid, 1991) is used for similar phenomena, describing the interaction of learning and identity-building in communities less depending on the institutional structure of firms but of the actual activities of professionals.

director with 36 years, feels almost too old for this job and art directors older than 40 are generally smiled at:

> The higher you come the narrower is the market, of course. And you cannot go back. As CD you cannot restart as art director or so [...]. When you meet old advertising professionals they are generally crazy. I have once worked with an art director from Frankfurt who was 45, but looked like 60. He appeared really broken. And he was crazy, really (personal interview, 2001).

As compared to other professions creative work appears to be depreciated with growing experience unless one has succeeded to advance in the professional career.

> The job is really ambivalent. On the one hand, it is really lively, you meet so many and interesting people, you participate, at least a little bit, in the world of glamour; you fly to exotic places, stay in the best hotels, and have dinner in the finest restaurants. On the other hand, it is awkward that professional experience does not really matter. When you are, say, a doctor, the older you are the more you can draw on your experience. That is a completely different self-confidence when you say: 'I am experienced.' In our job the older you get the more you lose self-confidence. And that is dangerous as it hinders you to do something against it (personal interview, 2001).

> Maybe it's therefore that you earn relatively much money in advertising; since you have to collect it in the first half of your life. Most of us do not do this until the normal age of retirement (personal interview, 2001).

Yet, although the problem of the short life span of the creative profession is evident and all professionals appear to be aware of it, there is neither any evidence of high unemployment rates for old creatives,[8] nor is this in any way raised as a problem in the sector discussions. It would be an interesting topic for an extra research project to follow the biographies of creatives older than 40 who have not been able to obtain management positions in advertising agencies focusing on the role of the interpersonal networks built up during the previous career to find a satisfying activity for the 'second half' of one's professional life.

Innovation as 'Factor Creation' and 'Factor Attraction' and the Functioning of a Professional Milieu: A Third Interpretation

In the course of the last three chapters we have – so to speak – 'screwed' our analysis continuously deeper into the logic of the advertising industry as well as of

---

8   Unemployment is a very rare phenomenon among creative advertising professionals in general. In June 2002 that is, in the middle of a severe crisis of the industry only 900 creatives were registered as unemployed throughout the whole of Germany. About 35 per cent of them were older than 39 years (unemployment statistics, own calculations).

its specific pattern in the German space-economy, having now finally reached the labour market as the crucial arena around which the functioning as well as the spatial organization of the industry is built. On the one hand, this appears self-evident given that ideas, that is, the output of advertising creatives, are the core which the success of the business relies on. On the other hand, one cannot isolate this core part from the wider context which it is involved in and which strongly influences the very performance of ideas creation. Methodologically, it has therefore been indispensable to draw the analytical bow as we have done, that is, so to speak from the general function and functioning of the industry to the biographies of individual workers.

Starting thus from portraying advertising as being involved in a complex set of relations both on the macro-level of the overall economy and on the micro-level of the single agency we pointed out that it was a significant restructuring of the whole system from the second half of the 1970s onwards which changed the framework of advertising also affecting its both on the global level and on the national level of the German territory where the advertising centre of gravity shifted from Frankfurt and Düsseldorf to Hamburg.

This phenomenon was conceived of as being based on 'window of locational opportunity' (Storper and Walker, 1989), a period of change in which new and innovative advertising agencies like Scholz & Friends and, above all, Springer & Jacoby, were able to 'produce' a new advertising region outside the traditional centres. In addition and as a first modification of the Californian school argument we stated that – unlike in the case of Storper and Walker's prototype of a region produced by an industry, Silicon Valley – the new German advertising region did not emerge as a greenfield development but within an existing metropolis.

In a second step we looked more closely at the nature of the innovation process in the context of a broader understanding of the advertising industry in terms of both its nature and its agency structure. We found that innovation basically signified to make the barrier between the two 'natures' of advertising, to be simultaneously a kind of popular art and a business service, more porous, thereby enhancing the status of the former. This innovation in turn restructured the agency landscape leading to a new pattern of competition above all on the labour market of creatives, being basically driven by the rivalry between the requirements of individual workers and the financial power of global advertising groups.

However, as we could see in the course of this chapter, the labour market itself has been an inherent element of innovation. Storper and Walker, as quoted several times, explain the relative locational freedom of first movers within a process of innovation by the fact that they are not bound to the constraints of mature industries in terms of factor inputs: 'Growing industries [...] enjoy [...] a factor-creating and factor-attracting power' (ibid., p.75). Given the evidence drawn from the German advertising industry, at least the factor creation can be held to be part of the innovation itself. A more artistic and entertaining type of advert needed new factors of production, that is, a new sort of advertising professional. Creating this advertising professional presupposed both getting access to labour force normally not accessed by a business service and adapting this artistic labour force to the needs of a business service.

It is at this point that the territorial dimension of innovation is touched: 'Factor-creation' in the sense of accessing particular groups of labour needs places where these groups can be accessed, as well as 'factor-attraction' needs places to where they can be attracted. It appears obvious that the new groups of potential employees coming from arts and literature were closely linked to urban lifestyles. Thus the locational freedom of the pioneers of German second wave advertising was only relative in that factor-creation and factor-attraction needed attractive metropolitan spaces. That it was precisely Hamburg was of course mainly owing to the biographies of the pioneers' founders but was also encouraged by the fact that the northern media centre offered both an attractive environment in terms of cultural life, architectural heritage etc. and the 'urban massivity' (Scott, 1997) of the second biggest city in Germany necessary for attracting young and talented people oriented to creative work.

Hence, 'producing' a new advertising region in the sense of Storper and Walker first of all meant producing a new advertising labour market within a given region that accomplished the preconditions for this labour market being established at all. Once established it developed effects of both spin-off and attraction leading to the increasing clustering of different types of advertising agencies in Hamburg. Spin-off, on the one hand, can be seen as a direct consequence of factor-creation and factor attraction in that the motivation of talented and ambitious labour force through the pioneer agencies simultaneously encouraged employees to become entrepreneurs. Attraction, on the other, was crucially owing to the new pattern of competition driven by the rivalry of individual desires and financial power. If (not only) global players intended to benefit from the innovation process they had to get access to the newly produced labour market. This in turn involved that they had both to get into the personalized networks which strongly determine this labour market and to deal with its volatile nature. Local clustering obviously underpins these networks thereby enabling agencies to handle volatility.

From the perspective of the individual creative professional the logic of agency clustering is similar: A variety of agencies offers a wide range of job opportunities that can be exploited as both trial-and-error field to find the right way for the individual professional biography and – more strategically – as steps to underpin the personal career promotion via 'lateral mobility'. In addition it helps to build personal ties that reduce risks and enhance opportunities in this context. Simultaneously these personal ties make up a 'community of equals', the professional milieu through which the assessment of an individual's activity is provided, through which common quality standards, conventions, but also narratives are produced and channelled through the labour market and through which each 'worker' is enabled to develop his or her individual identity within the context of an industry.

It is obvious that all these mechanisms being at work within a professional milieu are not absolutely limited to a particular territory. Narratives and conventions are also built through national and international events. Competition on the creative labour market occurs at least on a national level given that mobility of the mostly young creatives is relatively high. Thus global players do succeed to entice experienced creatives to other places. More recently Berlin has emerged as another 'massive' urban place to which both professionals can be attracted more easily and to which spin-offs increasingly tend to orient their activity. The decentralized structure of the German urban system appears to hinder the emergence of real exclusion effects in this context as compared, for instance, with the United States, France or the UK, where the main urban centres concentrate advertising employment in a much stronger way than it is the fact in Hamburg.

It is yet equally obvious that also in Germany local clustering in an urban environment strongly supports the functioning of a professional milieu by providing random opportunities for social relations to be built, by making rivalry between agencies transparent among the potential labour force as well as by fostering rivalry among the professionals themselves and by providing places at which the necessary sociability can be produced in everyday life.

This interaction between professional milieus and urban space leads us back to the conceptual issues of Chapter 2. On a general level it resembles the Marshallian argument of labour pooling in 'thick labour markets' in the light of an uncertain economic environment recently taken up by Krugman and others.[9] In addition, and more specifically, it reminds us of Camagni's recent further development of 'innovative milieus' within an urban context without ascribing to the city as a whole the quality of a milieu (Camagni, 1999).[10] The milieu in this sense offers the necessary cultural proximity between its members but also fosters the rivalry among them that drives its sustainable progress. Its functioning is yet underpinned by the urban context around it which both offers the members of the milieu to realize their lifestyles and guarantees the openness that provides the milieu with the necessary 'external energy' it needs to maintain innovativeness over the long run (Camagni, 1994, p.84).

Thus it seems to be the concurrence of an economic, a socio-cultural and a spatial logic (the latter in the sense of both spatial proximity and infrastructural richness) both innovation processes involving 'reflexive' labour and the structure and organization of 'reflexive' labour markets are based on. This brings us back to our general argument of considering precisely these labour markets as the key arenas through which the recontextualization of economic activities in a post-industrial knowledge economy occurs, in which thus our 'subject-oriented approach to understanding the interaction of economic organization, social relations and spatial structures'[11] is manifested. In the remainder of this book we shall reflect our findings on advertising in the light of this key argument, now deepening its spatial dimension only touched on at the end of Chapter 2.

---

9   See Chapter 2, p.37.

10 Ibid., p.23.

11 Ibid., p.39.

Chapter 6

Space and the Ambivalence of Reflexive Labour

Reflexive Labour and Space: Importance and Uncertainty

The role of labour for the spatial organization of industries already constituted a key element of the 'general' Weberian location theory. In this context labour was dealt with as an input into production, the cost of which – besides the transportation efforts for raw materials and final products to the consumer markets – was considered to be one of the variables which influence the location decision of manufacturing firms. This way of conceptualizing labour 'as just another "factor of production"', respectively as 'simple commodity' (Storper and Walker, 1989, p.154), has been heavily criticized in the research on industrial restructuring undertaken from the end of the 1970s onwards. Also Storper and Walker in this context emphasized the character of labour as a fictitious commodity, chiefly stressing the embeddedness of the labour force in structures of reproduction partly autonomous from labour demand (ibid., p.157). They argue that taking into account this 'unique character of labour supplies' would lead to the 'reinterpretation of the meaning of the term "spatial division of labour" as a social process rather than a managerial allocation of industrial facilities to predestined locations', thereby enabling academic work 'to push the labour "factor" to the forefront in our analysis of the geography of industrial capitalism' (ibid., p.154) and thus to resort to Weber's 'realist' or 'capitalist' theory of location which focuses on the nexus of a rising capitalist economy and the emergence of big urban labour markets.[1]

And indeed, there has been a great deal of work on industrial restructuring not only by the Californian School and other protagonists in Economic Geography of the 1970s and 1980s which both placed a greater emphasis on the role of labour in the process of reshaping the spatial organization of production and took a more comprehensive perspective than merely considering it as a 'factor of production'. In terms of our emphasis on knowledge-intensive or reflexive labour force roughly two main perspectives can be distinguished, this is, on the one hand, an interregional focus on the spatial division of labour along the overall process of industrial production, above all represented by Doreen Massey's work (Massey,

---

1   See Chapter 2, p.11.

1995 (1984)) and, on the other hand, a focus on the matching process within regional labour markets, respectively on the question how labour market mechanisms underpin the regional agglomeration of economic activities. This line of thought has strongly been pursued by Allen Scott in several studies on the labour market of animated film workers in Los Angeles (Scott, 1984b, 1988b), and it is close to the argument of 'labour pooling' already touched on at the ends of Chapter 2 and Chapter 5.

Massey's key point is that spatial division of labour constitutes one way of achieving the capitalists' double goal to increase both productivity and the degree of managerial control over production by, first, cutting the labour process into multiple steps of conceptualization and execution and, second, spreading it over space in order to exploit regional differentials of labour cost and/or unionization. In other words, spatial division of labour somehow constitutes an application of the principles of Taylorism to space. Similarly to Taylorism, however, it involves the problem of needing a strong investment in planning and control as well as in the technology enabling capitalists to manage the spreading of labour over space at all.[2] This in turn implies an opposite tendency to the 'deskilling' effect of the spatial division in that a growing body of highly skilled jobs is generated, mostly tending to concentrate in a restricted set of localities and attracting the corresponding industries to these places (Massey, 1995 (1984), p.137f.). Thus, Massey so to speak provides a spatially informed argument about the self-overriding dynamic inherent to Taylorism: The consequent implementation of its principles enhances the need for a particular reflexive labour input which tends to evade from capitalist control even being able to reverse the traditional territorial relation between the workforce and the working place, that is, 'that labour had to migrate to the places were the firms located' (Illeris, 1996, p.128).

Scott's argument, putting it in analogy to the nomenclature used to summarize Massey's points, is about spreading labour over time and not over space. Empirically it is based on the observation of extremely high turnover rates in the labour market of animated film workers – according to the cycle of television programming (Scott, 1984b, p.298). Given the impossibility of adjusting wages regular separation and re-accession is considered to be the only possibility for employers to deal with unstable demand in terms of labour cost. Basically they manage, by spreading labour over time according to its actual use, to avoid labour costs becoming fixed costs. In a second step Scott argues that high turnover rates 'encourage the spatial agglomeration of producers' (ibid., p.299) given that a large local labour market decreases turnover costs for both employers and employees since it lowers search and recruitment costs and offers employment alternatives for periods of layoff.

Interestingly this nexus of unstable labour markets and local agglomeration is particularly valid for the groups with the highest job status, that is, the creative or

---

2   See the brief discussion of Taylorism as a means to deal with the fictitious nature of labour as commodity in Chapter 2, pp.36f.

'above the line' employees.[3] In contrast to what can be drawn from Massey's argument on the clustering of highly skilled activities, reflexive labour in Scott's understanding is particularly subject to employers' cost minimization strategies.

Of course this is a problematic interpretation, and there are a lot of reasons for a fundamental critique of both approaches, particularly given that both undervalue the inherent socialness of labour force and labour markets. At least Massey's 'division of labour' argument does involve a social dimension, however similarly to the 'Global City'-debates discussed in Chapter 2,[4] it is merely conceptualized either as an outcome of economic mechanisms or as the power relations between capital and labour. In addition, regarding highly qualified, 'reflexive' labour, Massey exclusively focuses on labour's strong bargaining power which determines the organization of the corresponding industries. Nonetheless, despite these key shortcomings, Massey's and Scott's 'classic' work on the role of labour in spatial restructuring stresses a key ambivalence of reflexive labour that is, on the one hand, it has an increasingly important role as an input to production, reflected for instance in the high bargaining power of skilled workers and, on the other, it involves an increasing uncertainty regarding its outcome, so to speak as a 'radicalization of precisely the fictitious nature of labour in the light of an increasingly complex economy',[5] in Scott's argument manifested in the labour intensity of creative production and the risk of labour costs to become fixed costs.

This final chapter will start from both aspects, that is, first, from the critique of Massey's and Scott's argument as basically having a too structural perspective on the functioning of reflexive labour markets and, second, from the key ambivalence of importance and uncertainty. From this perspective it will attempt to unravel the spatial logic of reflexive labour markets, yet not as in the outlined approaches in form of a particular spatial outcome of functional-economic imperatives, as 'agglomeration' or 'location' of activities, but rather from the standpoint of the individualized subject both in work and life. Thus we shall now turn the common way of looking at the organization of the economy in space 'upside down' that is, we shall start our concluding discussion from the logic of the individual reflexive worker and confront it with the structural conditions each individual is involved in.

In analogy to the previous work we structure this along the 'trinity' of the economic, the social and the spatial which constituted a sort of 'latent' framework of both the theoretical discussion and the empirical analysis. Concerning the economic, the individual worker is considered to be primarily a *factor of production*, specifying more or less the ambivalence mirrored in the two outlined approaches; concerning the social it is above all regarded as *social being*, stressing thus the difficulties of integrating the professional dimension into a consistent

---

3 Similarly to the distinction of 'above-' and 'below-the-line' advertising also the work in creative industries can be broken down in creative 'above-the-line' and technical 'below-the-line' activities (Storper and Christopherson, 1987; Shapiro et al., 1992).

4 See pp.20ff.

5 See Chapter 2, p.37.

biography. The spatial organization – so our argument – is crucially shaped by the ambivalence of the *importance* of reflexive labour regarding both roles on one hand, and the multiple *uncertainties* inherent to them, on the other. In very general terms, space is held simultaneously to underscore the importance and to reflect the strategies of dealing with the uncertainties.

Yet this way of arguing from the standpoint of the individual worker implies a fundamental dilemma: on the one hand it aims at avoiding a premature attribution of specific features to particular groups – as it is done in most of the academic work dealing with the 'class' of post-industrial professionals – by focusing on the individual actor. On the other hand it is to come to generalizable conclusions regarding the mechanisms at work in the functioning of knowledge and culture intensive activities.

Despite being aware that the tension between the general and the particular will never be completely resolved in any kind of academic work in social science we believe that to explicitly interrelate individual agency with the structure it is involved in is a proper way not only to bypass the dilemma as well as possible but also to provide insights into the nature of this tension.

Reflexive Labour between the Economic, the Social and the Spatial

The Economic Logic: The Ambivalence of Knowledge-Labour as a Factor of Production

Simply put it is 'not just labour power' that is bought on labour markets of reflexive post-industrial professionals but 'knowledge' (Massey, 1995 (1984), p.138) incorporated in single individuals and only useable within them. Unlike forecasted by Daniel Bell the main feature of this knowledge is not its theoretic nature and its 'codification in abstract systems' (Bell, 1967, p.28) but – at least in terms of its application for economic purposes – its impossibility to be completely codified: a big portion of this knowledge remains 'tacit'.[6] In this sense knowledge-possession can be transferred from one person to another only to a limited extent, if at all, thereby constituting a sort of 'monopoly' (Massey, p.1995, p.138) which in turn – provided the knowledge can be exploited in business terms – endows the professional with a substantial bargaining power. The more specific and/or the less codifiable the knowledge-intensive activity of an employee is the higher – other things being equal – is his or her bargaining power. The more important knowledge-intensive work is for the functioning of an economic sector the more it is influenced by the bargaining power of the labour force. This is above all due to

---

6   The issue of tacit vs. codified knowledge has been widely discussed in the social science arena following to Daniel Bell, mainly drawing on Michael Polanyi's pioneering book 'The Tacit Dimension' (Polanyi, 1966). As a recent overview see Amin and Cohendet, 2004.

increased competition among firms which both aim to access the scarce good of specific knowledge and to exclude competitors from it thereby – as a side-effect of the bargaining power – enabling and encouraging employees to frequently change jobs (ibid.). In fact the advertising industry offers a whole range of examples of firms strongly competing chiefly for talented and successful creative labour force, thereby substantially fostering the labour mobility between firms.

This close nexus of knowledge-intensity and labour power and the self-reinforcing mechanisms of competition based on knowledge as production factor is overlaid and to a certain extent limited by two further elements:

First, knowledge-intensive labour is not only the decisive asset for the success of firms but at the same time by far the most important cost factor (Lash and Urry, 1994, p.199). This implies a fundamental trade-off which has to be resolved according to the specific situation of a firm. One can argue, as suggested by Massey, that cost-optimization occurs through productivity increases in the overall value chain, separating conception and execution and lowering the cost of the latter by innovation produced by the former. However, innovation can only rarely be conceived of as a linear process of productivity increase so that this argument may only apply to a part of knowledge-intensive activities. Thus, given the immunity of knowledge-work against productivity increases, it is crucial to avoid labour cost to become fixed cost by distributing it over time and/or fostering vertical disintegration as described by Scott. This in turn may result in an increasing virtualization of reflexive businesses thereby questioning the traditional borders of the firm.

Second, the less standardized, less codifiable the knowledge embodied in labour force is, the higher is not only labour's bargaining power. Also, with a decreasing degree of codification the uncertainty about the actual outcome of labour tends to grow, the more thus the fictitious nature of labour is radicalized. This aspect to a certain extent turns the common 'human capital'-based logic of the post-industrial knowledge economy upside down, since it challenges the nexus of higher qualification and increased competitiveness it is based upon at all. Knowledge has of course become the crucial element for economic activity, but the nature of knowledge-intensive labour has made business more complex given that the future profit earned by a worker cannot or only to a limited extent be a priori assessed. The strong efforts undertaken by firms in terms of human resources management, assessment centres etc. can largely be understood as means to reduce this uncertainty inherent to knowledge-intensive labour markets.

The creative labour of advertising examined in Chapters 3 to 5 constitutes an extreme case concerning all features of knowledge-intensity just outlined. The creation of 'ideas' lying at the heart of the advertising business, first, tends to be particularly immune to codification and standardization. Although there is a standard body of formats and rules for commercials and advertisements (Paczesny, 1988), and although successful ideas (in terms of creativity) tend to be imitated within the creative community, the very idea of communicating a particular product through the given set of media channels has to be unique for each case, a 'one-off', as it is the case basically in all culturally informed economic activities (Shapiro et al., 1992). Second, and closely linked to this, creative work tends to resist productivity increases. There have been, of course, drastic technological changes within advertising which have not only rationalized the interface between creation and print production but have also affected the very nature of creative work (Thinnes, 1996), however without being able to increase the productivity of ideas production. Third, and maybe most importantly, also the uncertainty concerning the future performance of labour force is even more accentuated in creative work given that it is not only the capacity embodied in the very personality of the professional but also the interplay between this capacity and the given situation within a firm the creative output is based on.

Thus, to conceive of an increasing knowledge- and/or culture-intensity of economic production in a wider sense as a mere rise of workers' bargaining power reaches too short. The bargaining power is indeed an important aspect but it is surrounded by a complex set of mechanisms at work, which entail a very conscious behaviour on the labour market on the part of both firms and professionals in order to reduce the increased uncertainty of the knowledge-intensive labour market. In addition, however, it reaches also too short to limit the discussion of the labour market to the narrow matching process of supply and demand thereby again to the aspect of 'static uncertainty' – even if in a wider sense as in Scott's model of transaction cost. Post-industrial labour as a factor of production is determined by (at least) two further aspects that above all display its involvement in a wider dynamic context:

First, knowledge embodied in workers is not just present but it is basically produced through formal education, training within firms etc. Supply of knowledge on the labour market thus presupposes to have successfully run through an institutional framework on the one hand 'transmitting industry-specific skills' (Christopherson and Storper, 1986), on the other hand providing individuals with the possibility to develop an own profile as specific advantage on the labour market.

Second, knowledge once acquired (along with experience obtained 'on the job') does not guarantee to be a life-long supply potential on the labour market. Given that specialized knowledge is not only characterized by non-codifiability and uncertainty regarding the economic success of this knowledge but also increasingly tends to be affected by short cycles of devaluation in the marketplace, it has to be constantly renewed in order to maintain bargaining power both within a firm and on the external market. The advertising case is again very specific and at the same time extreme in this context since not only the knowledge but the whole personality is devaluated once a certain age has been obtained.

Put in other words, the functioning of the post-industrial reflexive labour market is not only a matter of economic laws and imperatives but has to be seen as the interaction of the structural change of industries in which economically valuable knowledge is constantly restructured, on the one hand, and the professional biographies of individual professionals having to renew and reorganize their knowledge base, on the other. Continuing with our focus on the individual as starting point we shall thus continue discussing the latter, considering the fact to be a professional as part of a reflexive personal biography.

The Social Logic: The Dilemmas of a Professional 'Project of the Self'

As pointed out above,[7] there are in our view, two key explanations of why the contemporary economy has become 'more social', based on the concept of the reflexive subject as economic actor: First, seen from the standpoint of the single actor and drawing on and modifying Piore's economic reinterpretation of Hannah Arendt's concept of 'action', social embeddedness of economic activity is based on increasingly reflexive subjects' need for communities that acknowledge their economic activity thereby enabling them to 'give meaning to their lives' through professional action (Piore, 1990). Second, seen from the structural standpoint of Giddens' abstract 'expert systems', re-embedding of actually disembedded social systems occurs through 'facework commitments' within the expert systems themselves which need internal sociability to construct the identity of an inherently reflexive, that is, basically undefined and variable activity.

Professional biographies of knowledge-intensive labour force of course meet both aspects of embeddedness. Starting a career as professional means, first, choosing an activity which is compatible with the 'project of the self' (Giddens, 1991), with one's talents and interests, with one's attitude towards self-realization through the job at all etc. The process of finding this activity is not only rooted in the individual himself but occurs as interplay of a personal decision with the possibility to be confirmed or rejected, to succeed or to fail. The exemplary careers of two advertising art directors depicted in Chapter 5[8] have illustrated that the way into the career had by no means been self-evident but functioned as a process of trial and error shaped by the testing of own capabilities, acknowledgement by colleagues etc. Second, starting a career means to become part of an abstract expert system which develops its own professional 'ethos', its events, its conventions and its possibilities for sociability necessary for developing a common professional identity. Advertising, with its strong emphasis on 'events' within the professional community, be it the creativity contests, or agency football, volleyball or band championships, again constitutes an extreme example of such a professional facework logic within a particularly reflexive activity.

Engels (2002, pp.16f.) in a discussion of the changing ethos of professionalism in communication-oriented jobs to a certain extent further develops this line of thought, thereby offering a combination of these two perspectives of professional identity as part of the self-identity, on one hand, and part of the collective identity of a professional community, on the other. Basically she describes the entry into working life as a process of 'secondary socialization'. On the one hand, the

---

7 See Chapter 2, pp. 32f.
8 See pp.129ff.

importance of these secondary processes increases with the growing complexity of modern societies. On the other hand, they do not function with the same 'depth' as the intimate personal relations of primary socialization, but necessarily have the artificial character of pure knowledge transmission. The more open, the more undefined, that is, the more 'reflexive' an activity is, the less the artificial and standardized methods of professional socialization are able to construct a common ethos, in the sense of a both individual and collective identity of a professional community. A strong social underpinning of economic activities encompassing both the internal culture of enterprises and the wider professional milieu appears important to compensate the lack of such an ethos. The 'event culture' of advertising mirrors this kind of 'sociality' (Grabher, 2004, p.115), not based on strong ties of interpersonal relationship but on the combination of 'intensity' and 'ephemerality' (ibid.) thereby transmitting both a particular feeling of commonness and the industry's 'narratives' in which the criteria of performance assessment and reputation building are reproduced and the 'styles' and 'stars' are created. Of course there are also professional standards that make up a knowledge-intensive activity and that determine the career of advertising professionals as in our case. Yet, they would never be able to make the professional life part of a 'reflexive project of the self' (Giddens, 1991) constructed in the light of an exploding variety of choices.

But the dilemmas of a professional project of the self do not only involve the issue of a hybrid professional identity. Also, they are more closely linked to the fictitiousness of labour in the original sense, thus regarding the clash of the structural dynamic of an industry and the reflexive dynamic of the individual biography. In the case of advertising this conflict is particularly evident in the professionals' requirements as to their work atmosphere and the global groups' strategies of global reorganization, in that mergers change the management structure of one's agency thereby trying to sack those eventually responsible for the former bad performance. Yet it also embraces the more material interaction of work and life, regarding, for instance, the difficulties to balance family tasks with the time intensive project work, and the problem that after the age of 40 one's professionalism does no longer count. As the evidence has shown the networks of personal relations individual actors are tied in through their work serve as individual and collective management of the risks brought about by this interaction. Actually here is a self-reinforcing mechanism at work since the instability needs to be compensated by social relations but simultaneously produces them by driving people through different work contexts, thereby in turn fostering the instability since the number of contacts enhances the number of potential jobs.

In sum, also professional biographies of reflexive labour are characterized by a large degree of uncertainty which is either owing to the need for building identity within a reflexive activity or to the difficulty to choose the appropriate profession at all. In addition, and importantly, uncertainty emerges through the contact with the internal professional and structural economic logic of the abstract 'expert system' one enters when starting a professional career. It seems that reflexive activities provide a whole body of 'constellations' that make professional careers appear as having been the wrong way to have taken within the own reflexive project of the self.

The Spatial Logic: The Ambivalence of Importance and Uncertainty and the 'Local Dialectics of Labour'[9]

In a summarized view, the ambivalence of importance and uncertainty inherent to reflexive knowledge labour can be identified from both an economic and a social point of view. Concerning the economic the major significance of labour of course pushes firms to concentrate their strategy on getting and keeping the best people. This concentration is however strongly shaped by a series of uncertainties, both in a static sense of avoiding unnecessary fixed costs and in several dynamic senses of dealing with the increasingly tacit nature of knowledge etc. Concerning the social increasing reflexivity of individual professionals of course implies a strong importance of the professional identity for the individual biography. Given the diffuse character of reflexive activities, and given the high individual demands as to the 'success' of the biographical project it is yet highly uncertain whether also this positive integration of the professional into the biography succeeds. Put in other words, through the growth of reflexivity both economic activity and social life has become more sophisticated, but at the same time more difficult. This has been even intensified through the increasing 'convergence of work and life'[10] and the consequently growing degree of interaction of both spheres.

It is obvious that also the spatial dimension of this change towards reflexivity cannot be conceptualized as a straightforward spatial pattern of the knowledge economy. Neither does the focus on reflexive labour imply any clear interpretation of the secrets behind success or unsuccess of regions in comparison with others nor does it suggest the unanimous predominance of a particular spatial form such as the urban as compared to the non-urban. Thus, the assumption of a simple reverse of the traditional territorial nexus between workforce and workplace pointed out above obviously reaches too short. On the other hand, it is certainly not completely wrong given that knowledge-intensive labour markets – once existing – can function as a sort of attraction poles for economic activities as our advertising example has clearly shown for Hamburg within the context of the German space-economy.

In his fundamental book on 'the social regulation of labour markets' Jamie Peck (1996), basically drawing on labour market segmentation theory, holds that this social regulation implies an integral spatiality which is manifested on various scales of action. Thus, and obviously, national regulatory environments in terms of, for instance, the *'fait salarial'* (ibid., p.263, *original emphasis*) or other rules

---

9   Peck, 1996, p.261.
10 See Chapter 2, pp.30ff.

concerning the labour market generate a large degree of variance between the different spatial 'units' of nation states. Equally, also processes of 'labour mobility and market regulation' on the global level express the spatiality of the labour market. In addition to these more or less self-evident facts, which are also shared among academics not explicitly interested in the spatial structures of economic activity, he holds that 'the geography of labour markets is much more finely textured' (ibid.) offering two ways to understand the operation of labour markets on the local level, or – put in his own words – 'two sources' of the 'localness of the local labour market' (ibid., p.265), which he labels as 'production-reproduction dialectic', on the one hand, and as 'regulatory dialectic' on the other.

Thus Peck derives spatiality on the one hand from the matching of supply and demand on the labour market, through the fact that this process is inherently local and that it implies the embeddedness of the labour force in structures at least partly independent from the logic of production (Storper and Walker 1989, p.157), that is, the fact that it 'has to go home every night'. On the other hand labour markets are strongly shaped by their institutional underpinning and thus depend on their embeddedness in locally variable institutional and conventional environments. The former thus basically stresses the interplay between the individual and his or her integration into production, whereas the latter focuses more on the structural-functional logic of production itself.

Even if the term 'regulatory' resembles too much formal rules and even if Peck's book does not deal explicitly with the labour markets of post-industrial professionals, but instead tends to follow the predominating view of critical academics to discuss labour markets only from the standpoint of polarization and thus focuses on its 'lower end', these two dialectics constitute in our view useful starting points to discuss the spatial logic also of knowledge-intensive or 'reflexive' labour force. With his double focus on an individual and a functional-structural logic Peck also takes up the idea of seeing the spatial pattern as a particular field within the arena of the labour market in which individual 'reflexive' action interacts with the increasingly globalized 'structural' conditions it is embedded in, in which thus conflict between 'extensionality and intensionality' put forward by Giddens (1991) is settled. The subsequent final argumentation thus re-reads Peck's 'local dialectics of labour' along the topic of knowledge-intensity and reflexivity.

If we thus start with the 'production-reproduction dialectic' the logic appears to be – at first glance – very simple: Given that the importance of knowledge-intensive labour has increased, production – in terms of its spatial organization – has to seriously consider the desires or needs of the workforce regarding their reproduction sphere. The heightened bargaining power so to speak has a spatial dimension implying both a tendency of firms to move to areas 'where the people are' (Massey, 1995, p.137) – or a reverse of the traditional relation between workplace and living place – and the need for firms to take their employees' needs into account when choosing their locational strategy. In this sense the spatial dimension of knowledge-intensive labour can be discussed under the simple concept of 'soft location factors' (Grabow et al., 1995).

From the social point of view the logic is similar in that the border between the spheres of production and reproduction tend to blur. In current accounts this is generally labelled as a 'disenclosure of work and life' encompassing all 'social dimensions': time, space, technology, content/qualification, social organization and sense/motivation (Voß, 1998, p.479ff.).[11] Spatial implications might on the one hand be derived from the logic of time and the evidence drawn from the labour market of creatives in advertising suggests such an idea. The extreme strain through overtime work entails to spatially combine work and free-time, bringing about a relative proximity between the places of living, places of work and places of leisure, all of them located close to city centres. On the other hand, the spatial dimension can be directly affected by the approximation between the spheres of production and reproduction in that also the barriers between work place and living place tend to disappear through the emergence of telework, self-employed homework etc. (ibid.). The spatial 'outcome' can be both a dispersal and a concentration in city-centres, whichever the individual priorities of where private life shall 'take place' may be, or – put in other words – whichever location corresponds to the demands of an individual biographical project. The concentration of both advertising agencies and advertising staff in metropolitan centres is thus a result of the increasing *convergence of production and reproduction* given that the urban space meets the lifestyle of the creative labour force (Florida, 2002). The constraints set by frequent overtime work inherent to the project-oriented advertising work do certainly support this concentration, but apparently not as main vehicle.

As we argued above,[12] however, the clustering of activities around advertising functions also as a catalyst of the social 'milieu' relations which underpin the operation of the labour market in order to cope with its inherent risks and uncertainties. That is to say, Peck's 'production-reproduction dialectic' also has to be discussed in relation to the increased uncertainty linked with knowledge-intensive labour in terms of both its economic and its social logic. Clustering in this sense of course underpins social relations through which opportunities and risks are filtered. Yet it also enables both firms and employees to make errors relatively easily undone through providing them with alternative options. In this sense knowledge-intensive labour markets indeed require the pooling of specialized labour in the Marshallian sense as pointed out by Krugman (1991,

---

11 The 'tone' of such accounts frequently resembles the pessimistic positions complaining a subordination of all spheres of social life to an economic calculus (e.g. Sennett, 1998). An interesting interpretation is offered by Voß and Pongratz who describe the changes in the world of labour as a process in which social beings turn into 'labour force entrepreneurs' (Voß and Pongratz, 1998). However, they do not only reflect on this process as banalization or an 'emptying out' of subjects, but discover at least the 'latent civilizatory potentials' of the increased freedom inherent to this change (ibid., p.152).

12 See Chapter 5, pp.143f.

p.38ff.). Unlike in Marshall's and Krugman's interpretation this is however not so much based on unstable labour demand and increasing returns to scale, but rather on the risk of wrong choices and the need to avoid potential irreversibilities.

However, at least a part of the uncertainties appears to be too fundamental to be solved through mere economies of localization and also the recent reinterpretations of Marshallian 'thick labour markets' reach beyond the clustering of one industry (Glaeser, 1998). The fact that – seen from the perspective of individuals – the entry into working life is as little a self-evident and foreseeable operation as it is the continuity of a professional biography within one sector, as it can be said in a drastic way for the advertising 'age limit', involves the necessity of alternative job options outside the narrow sector but related to it.[13] For the example of advertising it is therefore that the existence of the wider field of media industries as a factor of location is of major importance. Seen from the firms' perspective – the existence of labour force outside the narrow field of advertising or any other knowledge-intensive activities allows for access to new segments, that is, for 'breaching traditionally closed spheres of exchange' which Granovetter (1990, p.18) describes as the key characteristic of 'entrepreneurship'. The fundamental innovations of the second wave of advertising can be discussed as such a 'breaking down of barriers' between different spheres of 'business' and 'popular arts' on the labour market. The more thus such innovations involve the labour market, the more they appear to depend on a large present variety of different segments. That is to say, dealing with the long-term uncertainty of long-term innovativeness ultimately requires the existence of strong economies of urbanization, yet not in the sense of the common use of general assets as infrastructures etc. but in terms of possible future innovations through hitherto unknown combinations of labour market spheres.[14] Agglomeration economies thus have to be strongly discussed in a socio-cultural sense of classic urban sociology, focusing on variety and 'difference' as key features of urban culture, which at the same time the cultural productivity of the city is based on (Robins, 1995; Scott, 1997).

The validity of the ambivalence of importance and uncertainty regarding the functioning of Peck's 'production-reproduction dialectic' on reflexive labour markets is continued in the second dimension of the 'localness of local labour markets', that is, the 'regulatory (or better: institutional) dialectic'. The crucial institutional issue in terms of knowledge activities appears to be the problem of 'skill-formation' (Peck, 1994, p.162ff.) given that it is precisely the intellectual capacities embodied in the labour-force that drive competitiveness. In a spatial perspective this would mean that different regional levels of workforce qualification imply a different degree of regional competitiveness. The crucial

---

13 Storper and Christopherson (1987, p.113), for the case of the motion picture industry, stress the vicinity to 'other entertainment industries' which require 'many of the skills demanded in motion picture industries'.

14 One can discuss this openness for future 'sources' of entrepreneurship also in the framework of Grannovetter's 'strength of weak ties'-theory (Granovetter, 1973, 1983).

institutions that provide knowledge skills can be said to be universities. That is to say, the existence of universities that produce both a large quantity and variety of well-qualified professionals provide regional firms with the necessary production factors thereby enhancing the overall competitiveness of the regional production system. From the social point of view universities additionally provide the socialization of individuals in professional communities necessary for the development of a professional identity. Continuing to use the terminology developed in the debates around economies of agglomeration universities can said to have the character of 'general assets', of public goods functioning as external economies of urbanization.

However, even though the increase of knowledge-intensity clearly correlates with an increase of the general level of formal qualification, also this way of knowledge-intensification is subject to strong uncertainties that shape the patterns of regional competitiveness, on several levels: First, as argued above, there is no linear relation between higher formal qualification and success with knowledge-intensive work. Most of the economically useable knowledge is both based on meta-competencies strongly linked to the personality of the labour force and built through an adaptation of formal skills to the practical needs of a business. Concerning the former the training of knowledge-intensive professionals comprises much more than mere academic education, thus entailing efforts in the whole range of (not only) educational activities. Concerning the latter, although the gap to be bridged for the entry into working life is not always as accentuated as between being (and feeling like) an artist and becoming an advertising professional, a certain degree of 'on-the-job' training is always necessary. In this context it is on one hand crucial for a firm's competitiveness whether or not the process of adaptation is successful. On the other hand, the enterprises' efforts conflict – at least to a certain extent – with the volatility of a knowledge-intensive labour market given that firms 'are unlikely to invest in the training of workers who are subsequently likely to leave' (ibid., p.161). In the case of advertising this dilemma has been resolved through a large degree of intra-sector co-operation at least in certain programmes such as the training of copywriters. Another possibility would be to relieve firms at least to a certain extent from the investment in training by building linkages between the institutional body of professional education and the firms, thus to breach also these barriers between universities and the region in order to 'defuse' the interface between training and work life. The problem of knowledge-labour formation, when discussed from a spatial perspective, is thus chiefly a matter of institutions, both from the very general standpoint of agglomeration economies guaranteeing a critical mass necessary for establishing good and big training institutions and from the more specific perspective of how these institutions can be made compatible with the demands of knowledge-intensive activities. This in turn depends both on an industry's capacity to common action and on political support, thus on strategic behaviour on the part of both entrepreneurs and policy-makers.

There is however a second dilemma that cannot be solved by targeted and strategic action alone, that is, knowledge has to be continuously reproduced and renewed both from the perspective of firms' competitiveness and from the single worker's point of view. On the contrary, too much targeted action runs the risk to become locked in strategic trajectories that do not allow for renewal.[15] Thus the performance of regional institutions and workers that pass through these institutions as well as through regional firms in a process of 'space-time filtering' (Scott, 1999b, p.47) depends on their openness to changes through external influences and through an existing variety of choices within the regions. Put in other words, also the continuous renewal of knowledge requires the socio-cultural 'variant' of urbanization economies: institutional and cultural diversity.

Table 6.1 The Spatial Logic of Reflexive Labour: A Basic Structure

Local Dialectics of Labour

		Production-Reproduction	Regulation (Institutions)
Ambivalence of Reflexivity	Importance	Rising significance of the reproduction sphere	Institutions for skill-formation (urbanization economies as general assets)
		Convergence of production and reproduction (Disenclosure of work and life)	
	Uncertainty	Specialized pooling of labour (localization economies)	Common strategic action to make skills useable
		Diversified pooling of labour (urbanization economies)	Institutional and cultural diversity to guarantee continuous renewal of skills

Source: Own illustration

Table 6.1 briefly summarizes our previously outlined arguments around the spatial logic of reflexive labour based on Jamie Peck's 'local dialectics of labour' and its confrontation with the ambivalence of importance and uncertainty inherent

---

15 See Grabher's (1993b) and others' arguments on the 'rigid specialization trap' outlined in Chapter 2, p.16f.

to reflexivity. It remarkably confirms that there is no linear way into the spatial forms of a reflexive economy. The upper two cells most strongly draw the idealized picture of knowledge, creativity etc. as key to competitiveness and the orientation of spatial organization along the reflexive knowledge labour as well as along the institutions that produce it. Massey's assertion that high-tech industries 'cluster [...] where the people are', added by a 'and where they are trained' as much represents this perspective as Richard Florida's idea of the economic success of regions strongly depending on their capacity to attract talented people. The two cells at the bottom however chiefly stress the uncertain nature of reflexive labour, focusing on the volatility of the overall employment system in terms of the lifetime of both single work contracts and the 'content' of work people have to adapt to. The spatial structures of a reflexive economy – in terms of economic actors' behaviour in space, of local policy etc. – will also be a reflection of, respectively a response to the increasing uncertainties inherent to a reflexive labour force.

Conceiving of the spatial structures of the knowledge economy thus requires to consider the fundamental ambivalence of reflexive labour, taking into account, on the one hand, the growing importance of knowledge workers and, on the other, the increasing uncertainties inherent to both the personalities and the work of these workers. Present accounts of future spatial organization and successful paths to regional competitiveness do not provide appropriate means to dealing with this ambivalence. To take the most popular examples: Neither the 'cluster' claim for a growing importance of localization economies as well as for their need to be institutionally underpinned (Porter, 1990) nor the 'node' view of strong 'fixed assets' to channel a global flow economy appear to be able to deal with the complexities of knowledge-intensity. Unlike them, the systematization in Table 6.1 suggests an increasing importance of economies of urbanization chiefly in a socio-cultural sense. A diverse socio-cultural context helps to avoid the potential irreversibility of wrong choices in an uncertain labour market both by firms and by individuals, by constantly providing alternative options for action, thereby both enabling actors to abandon wrong paths and constantly opening 'windows of opportunity' for the future. Thus the socio-cultural diversity and 'redundancy' (Grabher, 1994) of urban or metropolitan societies may play a major role in dealing with the inherently uncertain nature of post-industrial reflexive labour.[16] Even though there is by no means any uniform trend towards both a remedy for competitiveness and a single best form of spatial organization, big metropolitan regions – given both their diverse and complex structures and their institutional strength – are most likely to offer the 'reflexive environment' needed to underpin the performance of knowledge-intensive labour markets. The economic success of single regions may vary according to how this reflexive environment works, how it

16 This conclusion in its 'socially informed' view of the urban economy also differs from the most recent prominent account of the urban economy in which a rather modest view of cities is suggested, mainly conceiving of them in a merely institutional sense as 'sites' [...] 'in a distanciated economy' (Amin and Thrift, 2002, p.63ff.).

is able to produce and renew the regional skill base and to use it strategically without running the risk to be caught in the 'traps of rigid specialization'.

It is obvious that the structure of the table tends toward oversimplification particularly due to the multiplicity of uncertainties and due to the distinctive perspectives of economic and social logics. In addition it at least partly runs the risk to ignore the external mechanisms that drive a knowledge-intensive local or regional labour market. One could, for instance, add a further dialectic of 'global/local' which would be based on an increasing global competition for reflexive labour in the light of a future decreasing and ageing employment potential in the Western World. To discuss this more thoroughly would by far exceed the empirical material we were able to use in this book. Yet, it is obvious that also this fundamental change will, on the one hand, underpin the importance of knowledge-intensive labour for the functioning of the economy and, on the other hand, is likely to make this functioning more difficult, from the standpoint of both the performance of firms and of the workers' biographies. And there are many reasons to believe that these difficulties can be handled in the best way in the culturally diverse settings of big metropolitan regions.

Final Remarks and Critical Reflections

The present monograph aimed at providing a contribution to the manifold discussions on a recontextualization of economic activities in the post-industrial society, that is, on the increasing interrelatedness of the economy with the social and spatial environment it is embedded in. The both conceptual and empirical starting point was the so-called 'creative economy' which on one hand constitutes a paradigmatic example for this interrelatedness yet on the other suggests a particular focus in this context, that is, on the individual actor. Thus we basically followed a 'subject-oriented' or actor-centred approach to recontextualization, considering it as being mainly driven by and based on a 'convergence of work and life' on the level of the individual driven by the growing degree of 'reflexivity' which characterizes individual action in both spheres. The concept of 'reflexivity' draws mainly on Giddens' (1990, 1991) as well as Lash and Urry's (1994) work and means the increasing variety of options concerning all spheres of human life characteristic for the contemporary Western societies, implying to be set free from traditional structural constraints, but also to be confronted with a higher uncertainty about what is the right decision.

Starting from this general approach we argued that the identified convergence of work and life can by no means be considered as a harmonious merger but involves a series of conflicts rooted in this increasing uncertainty within each sphere as well as precisely in the closer interaction between them, given the essential difference regarding the basic criteria according to which individual action in both contexts is valued. The pivotal 'arena' in which this 'conflicting convergence' is manifested is the labour market of precisely the 'reflexive' professionals increasingly required to act independently from the 'rules and resources' of their work context, yet at the same time necessarily operating in a given context of material conditions set by increasingly global industries. The key argument of this book thus has been that it is the mechanisms within this labour market which chiefly shape the (not only) spatial organization of knowledge-intensive business activities.

This argument was confronted with a comprehensive empirical study into the spatial restructuring of the German advertising industry, portraying the shift of the national advertising centre from Frankfurt to Hamburg as a process of innovation within the framework of the industry's general change towards a 'second wave' characterized above all by a more creative and entertaining advertising style. This process set off by a group of newly established pioneer agencies chiefly involved the labour market of creative professionals by both accessing artistic labour market segments formerly not addressed by a business service and by adapting the creative capacities of this labour force to the needs of the advertising business. Innovation in advertising thus so to speak concerned a particularly pronounced case of a reflexive labour force. At the same time it displays a very good example of how the reflexive action of individuals necessary for an economic activity interacts and conflicts with its structural forces, on the one hand reshaping the pattern of competition in the sector thereby challenging the dominant role of big and hierarchically organized international advertising networks, on the other hand, yet driving these global players to react by making use of their higher financial power in order to enter the realm of creative advertising.

The interaction between the inherently reflexive action of creative professionals, necessary for a good performance in terms of the advertising standards but at the same time likely to involve a critical distance to the business, and the financial power of global communication groups also largely shapes the functioning of the labour market of creatives, implying a massive labour mobility between agencies, or between agency work and self-employment. Labour's high demands as to their work environment as well as active strategies of labour recruitment and labour buy-out, along with a creative activity's need for fresh air, encourage the high turnover rates. In turn the mobility is underpinned by well working networks of social relations, minimizing the risks inherent to high volatility and providing both firms and employees with a means to quickly react to changing labour demands and to strategically build careers while 'moving' through the agency landscape, respectively.

We finished our work concluding that the key function which labour markets of knowledge-intensive professionals have for the functioning and organization of the post-industrial economy does not mean that economic activities are simply oriented to the desires of their knowledge-intensive labour force. Given the multiplicity of conflicts concerning – roughly – the actual performance of 'reflexive' labour, the success of positively integrating professional life into a 'reflexive project of the self' (Giddens, 1991), or the intricate interaction of 'globalizing influences [...] and personal dispositions' (ibid.) the organization of knowledge labour is shaped by a fundamental 'ambivalence of importance and uncertainty' also and is strongly reflected in its spatial pattern. This spatial dimension is manifested in two dimensions: first, in the nexus of 'professional life' and 'social life' as necessarily territorialized relation of production and reproduction and, second, based on the 'production' of knowledge-intensive labour force in localized institutions. There are many aspects pointing to the fact that the multiplicity of overlapping logics encourages processes of metropolitanization, even though also this trend can by no means seen as linear given the diversity of choices about the 'design' of professional biographies.

At the moment there seem to be two important aspects regarding this argument about the spatiality of reflexive labour which appear under-explored and need to be further discussed as well as empirically scrutinized in a more profound way:

First, the convergence of production and reproduction and its impact on the spatial organization of knowledge-intensive activities also impinges both on markets which the internal organization of the urban economy and society is based upon, e.g. the housing market, the office market, the land market, and on the patterns of everyday life according to which people use urban space. It would be important to analytically transfer our arguments to both logics in order to lead the study of urban and regional economies from a network-focused analysis of structures and relations to a deeper examination of cities and regions which includes not only their economic performance but also their material form, the processes of exclusion and polarization produced by the interrelating dynamics of different urban markets and the patterns of daily work-life balance and city use (Jarvis, 2004). The overall objective of such an approach would be to combine the 'social' view of the urban economy proposed here with a perspective of 'cities as sites' and 'city rhythms' (Amin and Thrift, 2002), including a strong focus on intra-urban mechanisms.

Second, one can argue that the results so far presented are based too much on the extreme case of advertising. It would be wrong to jump from this evidence to the general conclusion about the spatial organization of the post-industrial economy. We would hold that of course the ideas presented here need further exploration also through case studies in less culturally informed activities. The initial idea to ascribe a pioneer function to advertising appears however legitimate given that the importance of culture-intensive products and services, or of products and services 'infused [...] with broadly aesthetic or semiotic attributes' (Scott, 1997, p.323) in the contemporary economy is still growing, tending to generalize a complicated convergence of culture and economy such as in advertising (ibid.; Lash and Urry, 1994; Crang, 1997, among many others). Consequently also the conflicting dynamics in increasingly reflexive labour markets are likely to be generalized. We have seen in the recent rise of the new media sector that the logic of creative labour found in advertising in terms of firm structures, spatial organization etc. has also become evident within a sector with a stronger technical orientation (Pratt, 2000; Scott, 1999b; Egan and Saxenian, 1999). Even though these innovations in the meantime have been experiencing a certain set-back in the wake of the dot.com crash one could maintain that the process initiated by advertising is now continuing to diffuse into the wider post-industrial economy.

In sum, the spatial organization of economic activities in the post-industrial era appears to reflect a full implementation of what Alfred Weber in his above mentioned, largely unnoticed 'capitalist theory of location'[17] called a 'labour-oriented industry' (Weber, 1923, p.76), in which, based on 'a decreasing level of transport cost and growing population densities' a 'complicated locational structure' can emerge (ibid.). According to Weber only the increasing disembedding of labour force through its 'mobilization' opens the way for this structure being really 'labour oriented', that is, driven by the 'motion laws' of the 'labour substratum of the modern capitalist era' (ibid.). This given he considers also the emergence of cities as mainly fostered by the growth of markets of commodified labour and the 'laws of flow and motion' that bring people into these markets.

Weber's account was strongly influenced by the massive rise of urbanization in the course of the industrialization processes in the late 19th and the early 20th century. Thus he basically described the same dynamics of disembedding as Karl Polanyi did in his concept of 'labour as a fictitious commodity' thereby stressing the role of the labour market as the essential link between economy and society. Although the function of labour as key input to the capitalist economy has changed significantly in the course of the last 150 years, this crucial link has remained and appears to be even more valid in the advanced period of modernity we are currently going through. And the lesson to be learnt from Weber is that it is this link the spatial organization of the economy is based on, in the 19th century as well as today. Thus, when discussing the role of reflexive (or, in our case study, creative) labour for the restructuring of the economy's spatial organization, encouraging to re-read Weber and to transfer his arguments to the present period of transformation appears to be both an appropriate conclusion and somehow a final demand for future work on the ongoing socio-economic change of cities and regions in the Western World.

---

17 See Chapter 2, p.11; Läpple, 1991.

Bibliography

Adorno, T. W. (1991), *The Culture Industry. Selected Essays on Mass Culture*, Routledge, London.

Advertising Age (2001), 'Agency Report 2001', *Advertising Age*, 17/2001.

AG Kulturwirtschaft (1995), *Culture and Media-Industries in the Regions of North Rhine-Westphalia*, 2nd Culture Industries Report, Summary, Ministerium für Wirtschaft, Mittelstand, Technologie und Verkehr des Landes Nordrhein-Westfalen, Düsseldorf.

Aksoy, A. and Robins, K. (1992), 'Hollywood for the 21st Century: Global Competition for Critical Mass in Image Markets', *Cambridge Journal of Economics*, Vol. 16, pp. 1-22.

Allen, J. (1999), 'Cities of Power and Influence: Settled Formations', in J. Allen, D. Massey and M. Pryke (eds), *Unsettling Cities: Movement/Settlement*, Routledge, London, pp. 181-218.

Amabile, T. A. (1997), 'Motivating Creativity in Organizations: On Doing What You Love and Loving What You Do', *California Management Review*, Vol. 40(1), pp. 39-58.

Amin, A. (2000), 'Industrial Districts', in E. Sheppard and T.J. Barnes (eds), *A Companion to Economic Geography*, Blackwell, Oxford, pp. 149-68.

Amin, A. and Cohendet, P. (2004), *Architectures of Knowledge: Firms, Capabilities, and Communities*, Oxford University Press, Oxford.

Amin, A. and Robins, K. (1991), 'These are not Marshallian Times', in R. Camagni (ed.), *Innovation Networks. Spatial Perspectives*, Belhaven Press, London, pp. 105-16.

Amin, A. and Thrift, N. (1992), 'Neo-Marshallian Nodes in Global Networks', *International Journal of Urban and Regional Research*, Vol. 16, pp. 571-87.

Amin, A. and Thrift, N. (2002), *Cities. Re-imagining the Urban*, Polity Press, Cambridge.

Arcelus, F. J. (1984), 'An extension of shift-share analysis', *Growth and Change*, 1/1984, pp. 3-8.

Arthur, W. B. (1994), *Increasing Returns and Path Dependence in the Economy*, University of Michigan Press, Ann Arbor.

Asheim, B. T. (1992), 'Flexible Specialization, Industrial Districts and Small Firms: A Critical Appraisal', in H. Ernste and V. Meier (eds), *Regional Development and Contemporary Industrial Response. Extending Flexible Specialization*, Belhaven Press, London, pp. 45-63.

Aydalot, P. and Keeble, D. (1988), 'High-Technology Industry and Innovative Environments in Europe: An Overview', in P. Aydalot and D. Keeble (eds), *High-technology Industry and Innovative Environments: The European Experience*, Routledge, London, pp. 1-21.

Bade, F.-J. and Niebuhr, A. (1999), 'Zur Stabilität des räumlichen Struktur-wandels', Jahrbuch für Regionalwissenschaft, Vol. 19, pp. 131-56.

Baethge, M. (2001), 'Qualifikationsentwicklung im Dienstleistungssektor', in M. Baethge and I. Wilkens (eds), Die große Hoffnung für das 21. Jahrhundert? Perspektiven und Strategien für die Entwicklung der Dienstleistungs-beschäftigung, Leske & Budrich, Opladen, pp. 85-106.

Baums, G. (1991), 'Die internationalen Agenturen auf dem Weg zur Triade', in J. Jeske, E. Neumann and W. Sprang (eds), Jahrbuch der Werbung in Deutschland, Österreich und der Schweiz, Vol. 28, Econ Verlag, Düsseldorf, Wien, New York, pp. 69-73.

Becattini, G. (1990), 'The Marshallian Industrial District as a Socio-Economic Notion', in F. Pyke, G. Becattini and W. Sengenberger (eds), Industrial Districts and Inter-Firm Co-operation in Italy, IILS, Geneva, pp. 37-51.

Beck, U. (1991), Risk Society: Towards a New Modernity, Sage, London.

Beck, U. (2000), The Brave New World of Work, Polity Press, Cambridge MA.

Beck, U., Giddens, A. and Lash, S. (1994), Reflexive Modernization, Polity Press, Cambridge MA.

Bell, D. (1974), The Coming of Post-Industrial Society. A Venture in Social Forecasting, Heinemann, London.

Berger, J. and Offe, C. (1984), 'Die Zukunft des Arbeitsmarktes: Zur Ergänzungsbedürftigkeit eines versagenden Allokationsprinzips', in C. Offe (ed.): 'Arbeitsgesellschaft'. Strukturprobleme und Zukunftsperspektiven, Campus, Frankfurt, pp. 87-117.

Bilton, C. and Leary, R. (2002), 'What Can Managers Do for Creativity? Brokering Creativity in the Creative Industries', International Journal of Cultural Policy, Vol. 8(1), pp. 49-64.

Block, F. (2001), 'Introduction', in K. Polanyi (2001 (1944)), The Great Transformation. The Political and Economic Origins of our Time, Beacon Press, Boston, pp. xviii-xxxviii.

Boden, D. and Molotch, H. (1994), 'The Compulsion of Proximity', in R. Friedland and D. Boden (eds), NowHere. Space, Time and Modernity, University of California Press, Berkeley, pp. 257-86.

Boldt, K. (1996), 'Das S&J-Phänomen. Deutschlands Paradewerber besinnen sich auf die alten Werte: einfach, einfallsreich, exakt', manager magazin, 9/96, pp. 105-12.

Boldt, K. (2001), 'Lob der Verführung. Kreativ-Index: Springer & Jacoby ist die kreativste Werbeagentur der Republik', manager magazin, 5/01, pp. 168-76.

Brito Henriques, E. and Thiel, J. (2000), 'The Cultural Economy of Cities: A Comparative Study of the Audiovisual Sector in Hamburg and Lisbon', European Urban and Regional Studies, Vol. 7(3), pp. 253-68.

Brown, J. S. and Duguid, P. (1991), 'Organizational Learning and Communities-of-practice: Toward a Unified View of Working, Learning, and Innovation', Organization Science, Vol. 2(1), pp. 40-57.

Brusco, S. (1990), 'The Idea of the Industrial District: Its Genesis', in F. Pyke, G. Becattini and W. Sengenberger (eds), Industrial Districts and Inter-Firm Co-operation in Italy, IILS, Geneva, pp. 10-19.

Buttler, F., Gerlach, K. and Liepmann, P. (1977), *Grundlagen der Regional-ökonomie*, Rowohlt, Reinbek bei Hamburg.

Camagni, R. (1991a), 'Introduction: From the Local "Milieu" to Innovation through Co-operation Networks', in R. Camagni (ed.), *Innovation Networks. Spatial Perspectives*, Belhaven Press, London, pp. 1-9.

Camagni, R. (1991b) 'Local "Milieu", Uncertainty and Innovation Networks: Towards a Dynamic Theory of Economic Space', in R. Camagni (ed.), *Innovation Networks. Spatial Perspectives*, Belhaven Press, London, pp. 121-44.

Camagni, R. (1994), 'Space Time in the Concept of "Milieu Innovateur"', in U. Blien, H. Herrmann and M. Koller (eds), *Regionalentwicklung und regionale Arbeitmarktpolitik. Konzepte zur Lösung regionaler Arbeitsmarktprobleme?*, Beiträge zur Arbeitsmarkt- und Berufsforschung 184, IAB, Nürnberg, pp. 74-89.

Camagni, R. (1999), 'The City as a Milieu: Applying the GREMI Approach to Urban Evolution', *Revue d'Économie Régionale et Urbaine*, 3/99, pp. 591-606.

Capecchi, V. (1990), 'A History of Flexible Specialization and Industrial Districts in Emilia-Romagna', in F. Pyke, G. Becattini and W. Sengenberger (eds), *Industrial Districts and Inter-Firm Co-operation in Italy*, IILS, Geneva, pp. 20-36.

Carnoy, M. (2000), *Sustaining the New Economy. Work, Family and Community in the Information Age*, Harvard University Press, Cambridge MA.

Castells, M. (1996), *The Rise of the Network Society*, The Information Age: Economy, Society and Culture, Vol.1, Blackwell, Malden.

Caves, R. E. (2000), *Creative Industries. Contracts between Art and Commerce*, Harvard University Press, Cambridge MA and London.

Chandler, A. D. (1977), *The Visible Hand. The Managerial Revolution in American Business*, Cambridge MA and London.

Chandler, A. D. (1992), 'What Is a Firm? A Historical Perspective', *European Economic Review*, Vol. 36, pp. 483-94.

Christopherson, S. and Storper, M. (1986), 'The City as Studio, the World as Back Lot: The Impacts of Vertical Disintegration on the Motion Picture Industry', *Environment and Planning D: Society and Space*, Vol. 3, pp. 305-20.

Crang, P. (1997), 'Introduction: Cultural Turns and the (Re)constitution of Economic Geography', in R. Lee and J. Wills (eds), *Geographies of Economies*, Arnold, London, pp. 3-15.

Daniels, P. (1995), 'The Internationalisation of Advertising Services in a Changing Regulatory Environment', *The Service Industries Journal*, Vol. 15(3), pp. 276-94.

David, P. A. (1985), 'Clio and the Economics of QWERTY', *American Economic Review*, Vol. 75(2), pp. 332-7.

David, P. A. (1994), 'Why are Institutions the "Carriers of History"?: Path Dependence and the Evolution of Conventions, Organizations and Institutions', *Structural Change and Economic Dynamics*, Vol. 5(2), pp. 205-20.

Dichtl, E. and Kaiser, A. (1981), 'Die Werbung in den Wirtschaftswissenschaften', in B. Tietz (ed.), *Die Werbung. Handbuch der Kommunikations- und*

Werbewirtschaft, Vol. 3, Verlag moderne Industrie, Landsberg am Lech, pp. 51-62.

Diekhof, R. (2001), 'Raus aus der Klassik-Gasse', *Werben und Verkaufen*, 26/2001, pp. 24-8.

Drake, G. (2003), '"This Place Gives me Space": Place and Creativity in the Creative Industries', *Geoforum*, Vol. 34(4), pp. 511-24.

Egan, E. A. and Saxenian, A. (1999), 'Becoming Digital. Sources of Localization in the Bay Area Multimedia Cluster', in H. J. Braczyk, G. Fuchs and H. G. Wolf (eds), *Multimedia and Regional Economic Restructuring*, Routledge, London, pp. 11-29.

Engels, K. (2002), 'Kommunikationsberufe im sozialen Wandel. Theoretische Überlegungen zur Veränderung institutioneller Strukturen erwerbsorientierter Kommunikationsarbeit', *Medien und Kommunikationswissenschaften*, Vol. 50(1), pp. 7-25.

Etzioni, A. (1988), *The Moral Dimension. Toward a New Economics*, The Free Press, New York.

Florida, R. (2002a), *The Rise of the Creative Class – and How It's Transforming Work, Leisure, Community and Everyday Life*, Basic Books, New York.

Florida, R. (2002b), 'The Geography of Bohemia', *Journal of Economic Geography*, Vol. 2, pp. 55-71.

Florida, R. (2003), 'Cities and the Creative Class', *City & Community*, Vol. 2(1), pp. 3-19.

Florida, R. and Gates, G. (2001), *Technology and Tolerance. The Importance of Diversity to High-Technology Growth*, The Brookings Institution Survey Series, June 2001.

Frank, B., Mundelius, M. and Naumann, M. (2004), 'Eine neue Geographie der IT- und Medienwirtschaft', *DIW-Wochenbericht*, 30/2004, pp. 433-40.

Geffken, M. (1999), 'Markenführung - die zentrale Aufgabe des Kommunikations- managements', in M. Geffken (ed.), *Das große Handbuch Werbung*, Verlag moderne Industrie, Landsberg am Lech, pp. 131-56.

Giddens, A. (1984), *The Constitution of Society. Outline of the Theory of Structuration*, University of California Press, Berkeley and Los Angeles.

Giddens, A. (1990), *The Consequences of Modernity*, Polity Press, Cambridge MA.

Giddens, A. (1991), *Modernity and Self-Identity*, Polity Press, Cambridge MA.

Gieseking, F. (2000), 'Wenn der Kunde fremdgeht', *Werben und Verkaufen*, 42/2000, pp. 122-24.

Gill, R. (2002), 'Cool, Creative and Egalitarian? Exploring Gender in Project Based New Media Work in Europe', *Information, Communication & Society*, Vol. 5(1), pp. 70-89.

Girard, M. and Stark, D. (2002), 'Distributing Intelligence and Organizing Diversity in New Media Projects', *Environment and Planning A*, Vol. 34, pp. 1927-49.

Glabus, W. (1991), 'Wettbewerb um Werbemillionen: wie große Unternehmen ihre Etats vergeben', in J. Jeske , E. Neumann, W. Sprang (eds), *Jahrbuch der Werbung in Deutschland, Österreich und der Schweiz*, Vol. 28, Econ Verlag, Düsseldorf, Wien, New York, pp. 65-8.

Glaeser, E. (1998), 'Are Cities Dying?', *Journal of Economic Perspectives*, Vol. 2, pp. 55-71.

Gonzalez, T. (2001), *Zwischenbetriebliche Verflechtungen im Druckgewerbe*, Unpublished Working Paper, TUHH, Hamburg.

Grabher, G. (1993a), 'Wachstumskoalitionen und Verhinderungsallianzen. Entwicklungsimpulse und Blockierungen durch regionale Netzwerke', *Informationen zur Raumentwicklung*, November, pp. 749-58.

Grabher, G. (1993b), 'The Weakness of Strong Ties. The Lock-in of Regional Development in the Ruhr Area', in G. Grabher (ed.), *The Embedded Firm: On the Socioeconomics of Industrial Networks*, Routledge, London, pp. 255-77.

Grabher, G. (1994), *Lob der Verschwendung. Redundanz in der Regionalentwicklung: Ein sozioökonomisches Plädoyer*, edition sigma, Berlin.

Grabher, G. (2001), 'Ecologies of Creativity: the Village, the Group and the Heterarchic Organisation of the British Advertising Industry', *Environment and Planning A*, Vol. 33, pp. 351-74.

Grabher, G. (2002), 'The Project Ecology of Advertising: Tasks, Talents and Teams', *Regional Studies*, Vol. 36(3), pp. 245-62.

Grabher, G. (2004), 'Learning in Projects, Remembering in Networks? Communality, Sociality, and Connectivity in Project Ecologies', *European Urban and Regional Studies*, Vol. 11(2), pp. 103-23.

Grabow, B., Henckel, D. and Hollbach-Grömig, B. (1995), *Weiche Standortfaktoren*, Schriften des Deutschen Instituts für Urbanistik 89, Kohlhammer, Stuttgart.

Granovetter, M. (1973), 'The Strength of Weak Ties', *American Journal of Sociology*, Vol. 78(6), pp. 1360-80.

Granovetter, M. (1983), 'The Strength of Weak Ties. A Network Theory Revisited', in R. Collins (ed.), *Sociological Theory*, Jossey-Bass, San Francisco, pp. 105-30.

Granovetter, M. (1985), 'Economic Action and Social Structure: The Problem of Embeddedness', *American Journal of Sociology*, Vol. 91(3), pp. 481-510.

Granovetter, M. (1990), *Entrepreneurship, Development and the Emergence of Firms*, Wissenschaftszentrum Berlin Discussion Papers FS I 90-2, WZB, Berlin.

Granovetter, M. (1992), 'The Sociological and Economic Approaches to Labour Market Analysis: A Social Structural View', in M. Granovetter and R. Swedberg (eds), *The Sociology of Economic Life*, Westview Press, Boulder and Oxford, pp. 233-63.

Habermas, J. (1981), *Theorie des kommunikativen Handelns*, Vol. 2, Suhrkamp Verlag, Frankfurt.

Hall, S. (1981), 'Encoding/Decoding', in S. Hall et al. (eds), *Culture, media, language. Working papers in cultural studies*, Routledge, London, pp. 128-38.

Harvey, D. (1989a), *The Condition of Postmodernity. An Enquiry into the Origins of Cultural Change*, Basil Blackwell, Oxford.

Harvey, D. (1989b), *The Urban Experience*, Basil Blackwell, Oxford.

Helbrecht, I. (1998), *The Creative Metropolis. Services, Symbols and Spaces*, Paper presented at the Jahrestagung der Gesellschaft für Kanada-Studien (GKS), Grainau.

Hochschild, A. R. (1997), *The Time Bind: When Work Becomes Home and Home Becomes Work*, Henry Holt, New York.

Howard, T. (1998), 'Survey of European Advertising Expenditure 1980-1996', *International Journal of Advertising*, Vol. 17, pp. 115-24.

Hudson, R. (1995), 'Making Music Work? Alternative Regeneration Strategies in a Deindustrialized Locality: The Case of Derwentside', *Transactions of the British Institute of Geographers*, Vol. 20, pp. 460-73.

Hutton, T. A. (2000), 'Reconstructed Production Landscapes in the Postmodern City: Applied Design and Creative Services in the Metropolitan Core', *Urban Geography*, Vol. 21(4), pp. 285-317.

Illeris, S. (1996), *The Service Economy. A Geographical Approach*, Wiley, Chichester.

Jackson, P. and Taylor, J. (1996), 'Geography and the Cultural Politics of Advertising', *Progress in Human Geography*, Vol. 20(3), pp. 356-71.

Jacoby, K. (1995), 'Ich bin schlichter Absatzförderer. Interview with Konstantin Jacoby' in in J. Kellner, U. Kurth and W. Lippert (eds), *1945 bis 1995: 50 Jahre Werbung in Deutschland*, Westermann, Ingelheim, pp. 126-9.

Jarvis, H. (2004), *City Time – Managing the Infrastructure of Everyday Life*, Paper presented at the ESRC Worklife Seminar Five, London, 27 February, 2004.

Jhally, S. (1987), *The Codes of Advertising. Fetishism and the Political Economy of Meaning in the Consumer Society*, Routledge, New York and London.

Jung, H. (1999), 'Medienexplosion', in M. Geffken (ed.), *Das große Handbuch Werbung*, Verlag moderne Industrie, Landsberg am Lech, pp. 39-49.

Kellner, J. (1995), '1945 bis 1995: 50 Jahre Entwicklung der Werbeagenturen in Deutschland', in J. Kellner, U. Kurth and W. Lippert (eds), *1945 bis 1995: 50 Jahre Werbung in Deutschland*, Westermann, Ingelheim, pp. 11-20.

Kemper, A. (2002), 'Interview with André Kemper, Managing Director of Springer & Jacoby', *new business*, 07/02/2002.

Kemper, A., Kröger, T. and Turner, S. (2000), 'Strategien für den Stern. Interview with André Kemper, Tonio Kröger, Sebastian Turner', *Werben und Verkaufen*, 36/2000, pp. 98-102.

Knox, P. and Agnew, J. (1994), *The Geography of the World Economy*, 2nd ed., Edward Arnold, London.

Kolle, S. (2002), 'Bleib wie Du bist. Verändere die Werbung. Interview with Stefan Kolle, Founder and Creative Director of Kolle Rebbe', in Die Zeit, GWP media-marketing (eds), *Unterbezahlt zu sein ist ein Supergefühl. Unbekannte Einblicke in Kopf und Bauch von Deutschlands Werbeelite*, Verlagsgruppe Handelsblatt, Hamburg.

Krätke, S. (2002), *Medienstadt: Urbane Cluster und globale Zentren der Kulturproduktion*, Leske & Budrich, Opladen.

Krätke, S. and Borst, R. (2000), *Berlin: Metropole zwischen Boom und Krise*, Leske & Budrich, Opladen.

Krugman, P. (1991), *Geography and Trade*, MIT Press, Cambridge MA.

Kunzmann, K. R. (1995), 'Strategien zur Förderung regionaler Kulturwirtschaft', in T. Heinze (ed.), *Kultur und Wirtschaft. Perspektiven gemeinsamer Innovationen*, Westdeutscher Verlag, Opladen, pp. 324-42.

Lange, H. J. (1999), 'Die Aufgaben der Werbung heute', in M. Geffken (ed.), *Das große Handbuch Werbung*, Verlag moderne Industrie, Landsberg am Lech, pp. 19-27.

Läpple, D. (1989), 'Zum Problem der ökonomischen Verflechtung – Das Beispiel der historischen Seehafenindustrien', in E. Pahl-Weber (ed.), *'Groß-Hamburg' nach 50 Jahren*, Verlag Dr. Krämer, Hamburg, pp. 35-60.

Läpple, D. (1991), 'Essay über den Raum. Für ein gesellschaftswissenschaftliches Raumkonzept', in H. Häußermann and D. Läpple (eds), *Stadt und Raum. Soziologische Analysen*, Centaurus, Pfaffenweiler, pp. 157-207.

Läpple, D. (1994), 'Zwischen gestern und übermorgen. Das Ruhrgebiet - eine Industrieregion im Umbruch', in R. Kreibich, A. S. Schmid, W. Siebel, T. Sieverts and P. Zlonicky (eds), *Bauplatz Zukunft. Dispute über die Entwicklung von Industrieregionen*, Klartext Verlag, Essen, pp. 37-51.

Läpple, D. (1998), 'Ökonomie der Stadt', in H. Häußermann (ed.), *Großstadt. Soziologische Stichworte*, Leske & Budrich, Opladen, pp. 193-207.

Läpple, D. and Kempf, B. (1999), *Die Entwicklung der Beschäftigung in der Freien und Hansestadt Hamburg. Kurzfassung des Zwischenberichts: Ausgewählte Ergebnisse der quantitativen Arbeitsmarktanalyse*, TUHH, Hamburg.

Läpple, D. and Kempf, B. (2001a), *Die Hamburger Arbeitslandschaft. Struktur und Entwicklung von Tätigkeitsfeldern im regionalen Vergleich. Ausgewählte Ergebnisse einer Arbeitsmarktstudie*, TUHH, Hamburg.

Läpple, D. and Kempf, B. (2001b), *Szenarien zur Hamburger Arbeitsmarktentwicklung in ausgewählten Tätigkeitsclustern. Kreative Tätigkeiten sowie Lager-, Transport- und Verkehrstätigkeiten*, TUHH, Hamburg.

Läpple, D. and Thiel, J. (1999), 'Unternehmensorientierte Dienstleistungen in westdeutschen Großstadtregionen – Hamburg im Vergleich', *Hamburg in Zahlen*, 3-4/99, pp. 41-52.

Lash, S. and Urry, J. (1994), *Economies of Signs and Space*, Sage, London.

Leborgne, D. and Lipietz, A. (1990), *Fallacies and Open Issues about Post-Fordism*, Centre d'Etudes Prospectives d'Economie Mathématique Appliquées à la Planification (CEPREMAP), Paris.

Leiss, W., Kline, S. and Jhally, S. (1990), *Social Communication in Advertising. Persons, Products and Images of Well-being*, 2nd ed., Nelson Canada, Scarborough, Ontario.

Leslie, D. A. (1995), 'Global Scan: The Globalization of Advertising Agencies, Concepts and Campaigns', *Economic Geography*, Vol. 71, pp. 402-26.

Leslie, D. A. (1997), 'Flexibly Specialized Agencies? Reflexivity, Identity, and the Advertising Industry', *Environment and Planning A*, Vol. 29, pp. 1017-38.

Leslie, D. A. (1999), 'Consumer Subjectivity, Space and Advertising Research' *Environment and Planning A*, Vol. 31, pp. 1443-57.

Leyshon, A. (1995), 'Annihilating Space?: The Speed-up of Communications', in J. Allen and C. Hamnett (eds), *A Shrinking World? Global Unevenness and Inequality*, Oxford University Press, Oxford and New York, pp. 11-46.

Maillat, D. and Vasserot, J. Y. (1988), 'Economic and Territorial Conditions for Indigenous Revival in Europe's Industrial Regions', in P. Aydalot and D. Keeble (eds), *High-technology Industry and Innovative Environments: the European Experience*, Routledge, London and New York, pp. 163-83.

Marshall, J. N. and Wood, P. (1995), *Services and Space. Key Aspects of Urban and Regional Development*, Longman, Harlow.

Martin, R. (2000), 'Institutional Approaches in Economic Geography', in E. Sheppard and T. J. Barnes (eds), *A companion to economic geography*, Blackwell, Oxford and Malden MA, pp. 77-94.

Massey, D. (1995 (1984)), *Spatial Division of Labour*, 2nd ed., Macmillan, London.

Mattelart, A. (1991), *Advertising International. The Privatization of Public Space*, Routledge, London.

Morgan, K. (1995), *Institutions, Innovation and Regional Renewal. The Development Agency as Animateur*, Paper Presented at the Regional Studies Association Conference 'Regional Futures: Past and Present, East and West', Gothenburg, 6-9 May, 1995.

Moulaert, F. and Djellal, F. (1995), 'Information Technology Consultancy Firms: Economies of Agglomeration from a Wide-Area Perspective', *Urban Studies*, Vol. 32(1), pp. 105-22.

Moulaert, F. and Lambooy. J. G. (1996), 'The Economic Organization of Cities: An Institutional Perspective', *International Journal of Urban and Regional Research*, Vol. 20(2), pp. 217-37.

Nelson, R. and Winter, S. (1982), *An Evolutionary Theory of Economic Change*, Harvard University Press, Cambridge MA.

Nov, O. and Jones, M. (2003), *Ordering Creativity? Knowledge, Creativity and Social Interaction in the Advertising Industry*, Proceedings of the 4th Conference on Organizational Knowledge, Learning and Capabilities, April, Barcelona.

NTC Publications (1998), *World Advertising Trends 1998*, NTC Publications, Oxfordshire.

Packard, V. (1957), *The Hidden Persuaders*, McKay, New York.

Paczesny, R. (1988), 'Was ist geheim and der Verführung? Strategien, Technik und Materialität der Werbung', in Gumbrecht, H. U. and Pfeiffer, K. L. (eds), *Materialität der Kommunikation*, Suhrkamp, Frankfurt, pp. 474-83.

Peck, J. (1994), 'Regulating Labour: The Social Regulation and Reproduction of Local Labour Markets', in A. Amin and N. Thrift (eds), *Globalization, Institutions and Regional Development in Europe*, Oxford University Press, Oxford, pp. 147-76.

Peck, J. (1996), *Work-Place. The Social Regulation of Labour Markets*, The Guilford Press, New York.

Peck, J. (2000) 'Places of Work', in E. Sheppard and T. J. Barnes (eds), *A Companion to Economic Geography*, Blackwell, Oxford, pp. 133-48.

Perry, M. (1990), 'The Internationalization of Advertising', *Geoforum*, Vol. 21(1), pp. 35-50.

Piore, M. (1990) 'Work, Labour and Action: Work Experience in a System of Flexible Production', in F. Pyke, G. Becattini and W. Sengenberger (eds), *Industrial Districts and Inter-Firm Co-operation in Italy*, IILS, Geneva, pp. 52-74.

Piore, M. and Sabel, C. F. (1984), *The Second Industrial Divide. Possibilities for Prosperity*, Basic Books, New York.

Pohlan, J. (2001), 'Monitoring der Städte und Regionen', in N. Gestring, H. Glasauer, C. Hannemann, W. Petrowsky and J. Pohlan (eds), *Jahrbuch StadtRegion 2001. Schwerpunkt: Die Einwanderungsstadt*, Leske + Budrich, Leverkusen, pp. 205-59.

Polanyi, K. (2001 (1944)), *The Great Transformation. The Political and Economic Origins of our Time*, Beacon Press, Boston.

Polanyi, M. (1966), *The Tacit Dimension*, Doubleday, Garden City N.Y.

Porter, M. (1990), *The Competitive Advantage of Nations*, Macmillan, London.

Powell, W. W. and Smith-Doerr, L. (1994), 'Networks and Economic Life', in N. J. Smelser and R. Swedberg (eds), *The Handbook of Economic Sociology*, Princeton University Press, Princeton, pp. 368-402.

Pratt, A. C. (2000), 'New Media, the New Economy and New Spaces', *Geoforum*, Vol. 31(4), pp. 425-36.

Pratt, A. C. (2004), 'The Cultural Economy. A Call for Spatialized "Production of Culture" Perspectives', *International Journal of Cultural Studies*, Vol. 7(1), pp. 117-28.

Putnam, R. D. (2000), *Bowling Alone: The Collapse and Revival of American Community*, Simon and Schuster, New York.

Richter, K. and Peus, A. (1999), 'Des Werbers Lohn. Preisgekrönte Kampagnen aus Deutschland in nationalen und internationalen Wettbewerben', in W. Schalk, H. Thoma and P. Strahlendorf (eds), *Jahrbuch der Werbung für den deutschsprachigen Raum*, Vol. 36, Econ-Verlag, München und Düsseldorf, pp. 116-25.

Robins, K. (1995), 'Collective Emotion and Urban Culture', in P. Healey, S. Cameron, S. Davoudi, S. Graham and A. Madani (eds), *Managing Cities. The New Urban Context*, Wiley, Chichester, pp. 45-61.

Sabel, C. F. (1989), 'Flexible Specialization and the Re-Emergence of Regional Economies', in P. Hirst and J. Zeitlin (eds), *Reversing Industrial Decline*, Berg, Oxford, pp. 17-70.

Sassen, S. (1991), *The Global City. New York, London, Tokyo*, Princeton University Press, Princeton NJ.

Sassen, S. (1994), *Cities in a World Economy*, Pine Forge Press, Thousand Oaks.

Sassen, S. (1995), 'On Concentration and Centrality in the Global City', in P. L. Knox and P. J. Taylor (eds), *World Cities in a World-System*, Cambridge University Press, Cambridge, pp. 63-75.

Sassen, S. (2001), 'Global Cities and Global City-Regions: A Comparison', in A. J. Scott (ed.), *Global City-Regions. Trends, Theory, Policy*, Oxford University Press, Oxford, pp. 78-95.

Saxenian, A. (1994a), *Regional Advantage. Culture and Competition in Silicon Valley and Route 128*, Harvard University Press, Cambridge MA.

Saxenian, A. (1994b), 'Lessons from Silicon Valley', *Technology Review*, July, pp. 42-51.

Saxenian, A. (2001), 'The Role of Immigrant Entrepreneurs in New Venture Creation', in C. B. Schoonhoven and E. Romanelli (eds), *The Entrepreneurship Dynamic: Origins of Entrepreneurship and the Evolution of Industries*, Stanford University Press, Stanford, pp. 68-108.

Sayer, A. (1997), 'The Dialectic of Culture and Economy', in R. Lee and J. Wills (eds), *Geographies of Economies*, Arnold, London, pp. 16-26.

Sayer, A. and Walker, R. (1992), *The New Social Economy. Reworking the Division of Labour*, Blackwell, Cambridge MA and Oxford.

Scheffler, H. (1999), 'Zielgruppenvielfalt', in M. Geffken (ed.), *Das große Handbuch Werbung*, Verlag moderne Industrie, Landsberg am Lech, pp. 31-7.

Schmidt, S. J. (1995), 'Werbung zwischen Wirtschaft und Kunst', in S. J. Schmidt and B. Spieß (eds), *Werbung, Medien und Kultur*, Westdeutscher Verlag, Opladen, pp. 26-43.

Schmidt, S. J. (1999), 'Werbung', in J. Wilke (ed.), *Mediengeschichte der Bundesrepublik Deutschland*, Bundeszentrale für politische Bildung, Bonn, pp. 518-36.

Schmidt, S. J. and Spieß, B. (1994), *Die Geburt der schönen Bilder. Fernsehwerbung aus Sicht der Kreativen*, Westdeutscher Verlag, Opladen.

Schmidt, S. J. and Spieß, B. (1997), *Die Kommerzialisierung der Kommunikation. Fernsehwerbung und sozialer Wandel 1956-1989*, Suhrkamp, Frankfurt.

Schoenberger, E. (1997), *The Cultural Crisis of the Firm*, Blackwell, Cambridge MA and Oxford.

Schoenberger, E. (2000), 'Creating the Corporate World: Strategy and Culture, Time and Space', in E. Sheppard and T. J. Barnes (eds), *A Companion to Economic Geography*, Blackwell, Oxford, pp. 377-91.

Scholz, J. (1998), 'Jürgen Scholz. Bilanz eines Werbe-Lebens. Interview with Jürgen Scholz', in Grimm, R. (ed), *Das Medium - 50 Jahre Werbung im Stern*, Gruner + Jahr, Hamburg, pp. 60-67.

Scholz & Friends (2001), 'Scholz & Friends Group GmbH und United Visions Entertainment AG gehen zusammen', *Press Release*, http://www.scholz-and-friends.com/deutsch/home/special2.html.

Scholz & Friends (2004), *Scholz & Friends Annual Report 2003*, Scholz & Friends, Berlin.

Schröter, H. G. (1997), 'Die Amerikanisierung der Werbung in der Bundesrepublik Deutschland', *Jahrbuch für Wirtschaftsgeschichte*, 1997/1, pp. 93-115.

Schumpeter, J. A. (1988a (1947)), 'The Creative Response in Economic History', in R. V. Clemence (ed.), *Joseph A. Schumpeter. Essays on Entrepreneurs, Innovations, Business Cycles, and the Evolution of Capitalism*, Transaction Publishers, New Brunswick and London, pp. 221-31.

Schumpeter, J. A. (1988b (1949)), 'Economic Theory and Entrepreneurial History', in R. V. Clemence (ed.), *Joseph A. Schumpeter. Essays on*

Entrepreneurs, Innovations, Business Cycles, and the Evolution of Capitalism, Transaction Publishers, New Brunswick and London, pp. 253-71.

Scott, A. J. (1984a), 'Industrial Organization and the Logic of Intra-Metropolitan Location II: A Case Study of the Women's Dress Industry in the Greater Los Angeles Region', *Economic Geography*, Vol. 60, pp. 3-27.

Scott, A. J. (1984b), 'Territorial Reproduction and Transformation in a Local Labour Market: The Animated Film Workers of Los Angeles', *Environment and Planning D: Society and Space*, Vol. 2, pp. 277-307.

Scott, A. J. (1988a), *New Industrial Spaces*, Pion, London.

Scott, A. J. (1988b), *Metropolis. From the Division of Labour to Urban Form*, University of California Press, Berkeley.

Scott, A. J. (1996), 'The Craft, Fashion, and Cultural Products Industries of Los Angeles: Competition Dynamics and Policy Dilemmas in a Multisectoral Image-Producing Complex', *Annals of the Association of American Geographers*, Vol. 86(2), pp. 306-23.

Scott, A. J. (1997), 'The Cultural Economy of Cities', *International Journal of Urban and Regional Research*, Vol. 21(2), pp. 323-40.

Scott, A. J. (1999a), 'The Cultural Economy: Geography and the Creative Field', *Media, Culture & Society*, Vol. 21(6), pp. 807-17.

Scott, A. J. (1999b), 'Patterns of Employment in Southern California's Multimedia and Digital Visual-Effects Industry. The Form and Logic of an Emerging Local Labour Market', in H.J. Braczyk, G. Fuchs and H. G. Wolf (eds), *Multimedia and Regional Economic Restructuring*, Routledge, London, pp. 30-48.

Scott, A. J. (2000), 'French Cinema. Economy, Policy and Place in the Making of a Cultural Products Industry' *Theory, Culture & Society*, Vol. 17(1), pp. 1-38.

Sennett, R. (1998), *The Corrosion of Character*, W.W. Norton, New York.

Shapiro, D., Abercrombie, N., Lash, S. and Lury, C. (1992), 'Flexible Specialization in the Culture Industries', in H. Ernste and V. Meier (eds), *Regional Development and Contemporary Industrial Response. Extending Flexible Specialization*, Belhaven Press, London and New York, pp. 179-94.

Storper, M. (1995), 'The Resurgence of Regional Economies, Ten Years Later. The Region as a Nexus of Untraded Interdependencies', *European Urban and Regional Studies*, Vol. 2(3), pp. 191-221.

Storper, M. (1997), *The Regional World. Territorial Development in the Global Economy*, The Guilford Press, New York.

Storper, M. and Christopherson, S. (1987), 'Flexible Specialization and Regional Industrial Agglomerations: The Case of the U.S. Motion Picture Industries', *Annals of the Association of American Geographers*, Vol. 77, pp. 104-17.

Storper, M. and Scott, A. J. (1995), 'The Wealth of Regions. Market Forces and Policy Imperatives in Local and Global Context', *Futures*, Vol. 27(5), pp. 505-26.

Storper, M. and Walker, R. (1989), *The Capitalist Imperative: Territory, Technology and Industrial Growth*, Basil Blackwell, New York.

Strahlendorf, P. (1999), 'Die 'Newcomer-Agentur des Jahres 1998' heißt weigertpirouzwolf in Hamburg' in W. Schalk, H. Thoma and P. Strahlendorf (eds), *Jahrbuch der Werbung für den deutschsprachigen Raum*, Vol. 36, Econ-Verlag, München und Düsseldorf, pp. 14-18.

Swedberg, R. and Granovetter, M. (1992), 'Introduction', in M. Granovetter and R. Swedberg (eds), *The Sociology of Economic Life*, Westview Press, Boulder and Oxford, pp. 1-26.

Taylor, P., Walker. D. R. F., Catalano, G. and Hoyler, M. (2002), 'Diversity and Power in the World City Network', *Cities*, Vol. 19(4), pp. 231-41.

Thinnes, P. (1996), *Arbeitszeitmuster in Dienstleistungsbetrieben. Eine zeit- und organisationssoziologische Untersuchung am Beispiel der Werbebranche*, Campus, Frankfurt and New York.

Thrift, N. (1987), 'The Fixers: the Urban Geography of International Commercial Capital', in J. Henderson and M. Castells (eds), *Global Restructuring and Territorial Development*, Sage, London, pp. 203-33.

Thompson, P. and Warhurst, C. (2004), *New Forms of Work and Organization in the Creative Industries*, Paper Presented at the International Workshop 'Studying New Forms of Work. Concepts and Practices in Cultural Industries and Beyond', Free University of Berlin, 26-27 March, 2004.

Touraine, A. (2002), *Old and New Social Models*, The 9th Hamburg Lecture, 28 January 2002, unpublished manuscript.

Veltz, P. (1997), 'The Dynamics of Production Systems, Territories and Cities', in F. Moulaert and A. J. Scott (eds), *Cities, Enterprises and Society on the Eve of the 21st Century*, Pinter, London, pp. 78-96.

Von Matt, J. R. (1995), 'In der Außenwerbung steckt das größte Innovations-potential. Interview with Jean-Remy von Matt', in J. Kellner, U. Kurth, W. Lippert (eds), *1945 bis 1995: 50 Jahre Werbung in Deutschland*, Westermann, Ingelheim, pp. 148-51.

Von Matt, J. R. (2001), 'Zwischen Sein und Schein. Interview with Jean-Remy von Matt', *Werben und Verkaufen*, 25/2001, pp. 24-8.

Von Matt, J. R. (2002), 'Unterbezahlt zu sein ist ein Supergefühl. Interview with Jean-Remy von Matt', in Die Zeit, GWP media-marketing (eds), *Unterbezahlt zu sein ist ein Supergefühl. Unbekannte Einblicke in Kopf und Bauch von Deutschlands Werbeelite*, Verlagsgruppe Handelsblatt, Hamburg.

Voß, G. G. (1998), 'Die Entgrenzung von Arbeit und Arbeitskraft. Eine subjekt-orientierte Interpretation des Wandels der Arbeit', *Mitteilungen aus der Arbeitsmarkt- und Berufsforschung*, Vol. 31(3), pp. 473-87.

Voß, G, G, and Pongratz, H. J. (1998), 'Der Arbeitskraftunternehmer. Eine neue Grundform der Ware Arbeitskraft?', *Kölner Zeitschrift für Soziologie und Sozialpsychologie*, Vol. 50(1), pp. 131-58.

Weber, A. (1923), 'Industrielle Standortslehre (Allgemeine und kapitalistische Theorie des Standorts)', in *Grundriss der Sozialökonomik VI. Abteilung: Industrie, Bauwesen, Bergwesen*, J. C. B. Mohr, Tübingen, pp. 58-86.

Webster, F. and Robins, K. (1989), 'Plan and Control. Towards a Cultural History of the Information Society', *Theory and Society*, Vol. 18, pp. 323-51.

Wells, W., Burnett, J. and Moriarty, S. E. (1989), *Advertising. Principles and Practice*, Prentice-Hall, Englewood Cliffs.

Willenbrock, H. (2000), 'Eine Idee vom Leben. St. Luke's', *brand eins,* No. 9.

Wischermann, C. (1995a), 'Einleitung: Der kulturgeschichtliche Ort der Werbung', in P. Borscheid and C. Wischermann (eds), *Bilderwelt des Alltags. Werbung in der Konsumgesellschaft des 19. und 20. Jahrhunderts*, Franz Steiner Verlag, Stuttgart, pp. 8-19.

Wischermann, C. (1995b), 'Grenzenlose Werbung? Zur Ethik der Konsumgesellschaft', in P. Borscheid and C. Wischermann (eds), *Bilderwelt des Alltags. Werbung in der Konsumgesellschaft des 19. und 20. Jahrhunderts*, Franz Steiner Verlag, Stuttgart, pp. 372-407.

Wolfe, M. R. (1999), 'The Wired Loft. Lifestyle Innovation Diffusion and Industrial Networking in the Rise of San Francisco's Multimedia Gulch', *Urban Affairs Review, Vol.* 34(5), pp. 707-28.

ZAW/Zentralverband der deutschen Werbewirtschaft (1997), *Werbung in Deutschland 1997*, Verlag edition ZAW, Bonn.

ZAW/Zentralverband der deutschen Werbewirtschaft (2001), *Werbung in Deutschland 2001*, Verlag edition ZAW, Bonn.

ZAW/Zentralverband der deutschen Werbewirtschaft (2002), *Werbung in Deutschland 2002*, Verlag edition ZAW, Bonn.

Ziegler, F. (1994), *Internationale Wettbewerbsfähigkeit von Dienstleistungsbranchen. Eine empirische Analyse der Werbebranche,* Peter Lang, Bern et al.

Zuberbier, I. (1981), 'Die Werbeagentur – Funktionen und Arbeitsweise', in B. Tietz (ed.), *Die Werbung. Handbuch der Kommunikations- und Werbewirtschaft*, Vol. 3, Verlag moderne Industrie, Landsberg am Lech, pp. 2373-406.

Zukin, S. (1995), *The Cultures of Cities*, Blackwell, Cambridge MA.

Index

Milton Keynes UK
Ingram Content Group UK Ltd.
UKHW031150141024
449569UK00024B/915

9 781138 619203